Predator-Prey Relationships

Photo by Gerhard Roth. Used with permission of Mr. Roth and the *Journal of Morphology.*

Predator-Prey Relationships

Perspectives and Approaches
from the Study of Lower Vertebrates

Edited by
Martin E. Feder and George V. Lauder

The University of Chicago Press
Chicago and London

MARTIN E. FEDER is in the Department of Anatomy of the University of Chicago, and GEORGE V. LAUDER is in the School of Biological Sciences of the University of California, Irvine.

The University of Chicago Press, Chicago 60637
The University of Chicago Press, Ltd., London
© 1986 by The University of Chicago
All rights reserved. Published 1986
Printed in the United States of America
95 94 93 92 91 90 89 88 87 86   5 4 3 2 1

Library of Congress Cataloging-in-Publication Data

Main entry under title:
Predator-prey relationships.

   Includes index.
   1. Predation (Biology)—Congresses.   I. Feder,
Martin E.   II. Lauder, George V.   III. Title: Lower
vertebrates.
QL758.P72   1986   597'.0524   85-24709
ISBN 0-226-23945-4
ISBN 0-226-23946-2 (pbk.)

Contents

Contributors   vii

Preface   ix

1. Introduction   1
    George V. Lauder and Martin E. Feder

2. Functional Morphology of   6
    Predator-Prey Relationships
    Carl Gans

3. Locomotion and Predator-Prey Relationships   24
    Paul W. Webb

4. Neural Mechanisms of Prey Recognition: An   42
    Example in Amphibians
    Gerhard Roth

5. Measuring Behavioral Energetics   69
    Albert F. Bennett

6. A Comparative Approach to Field and Laboratory   82
      Studies in Evolutionary Biology
   Raymond B. Huey and Albert F. Bennett

7. Natural History and Evolutionary Biology   99
   Harry W. Greene

8. Defense Against Predators   109
   John A. Endler

9. Behavioral Responses of Prey Fishes During   135
      Predator-Prey Interactions
   Gene S. Helfman

10. Laboratory and Field Approaches to the   157
      Study of Adaptation
   Stevan J. Arnold

11. Commentary and Conclusion   180
   Martin E. Feder and George V. Lauder

Index   191

Contributors

Stevan J. Arnold, *Department of Biology, University of Chicago, 940 East 57th Street, Chicago, Illinois 60637*

Albert F. Bennett, *School of Biological Sciences, University of California, Irvine, California 92717*

John A. Endler, *Department of Biology, University of Utah, Salt Lake City, Utah 84112*

Martin E. Feder, *Department of Anatomy, University of Chicago, 1025 East 57th Street, Chicago, Illinois 60637*

Carl Gans, *Division of Biological Sciences, University of Michigan, 2127 Natural Sciences Building, Ann Arbor, Michigan 48109*

Harry W. Greene, *Museum of Vertebrate Zoology and Department of Zoology, University of California, Berkeley, California 94720*

Gene S. Helfman, *Department of Zoology and Institute of Ecology, University of Georgia, Athens, Georgia 30602*

Raymond B. Huey, *Department of Zoology NJ-15, University of Washington, Seattle, Washington 98195*

George V. Lauder, *Department of Anatomy, University of Chicago, 1025 East 57th Street, Chicago, Illinois 60637; Present address: School of Biological Sciences, University of California, Irvine, California 92717*

Gerhard Roth, *Department of Biology, University of Bremen, Bremen, Federal Republic of Germany*

Paul W. Webb, *School of Natural Resources, University of Michigan, Ann Arbor, Michigan 48109*

Preface

This book is an outgrowth of a symposium held in 1985 at the joint meeting of the American Society of Ichthyologists and Herpetologists, Herpetologists' League, and Society for the Study of Amphibians and Reptiles in Norman, Oklahoma. When we planned the symposium and invited the speakers, we made no provision for publication and did not anticipate the consensus that developed during and shortly after the symposium that a published volume would be a valuable addition to the literature.

In both organizing the symposium and in compiling this book, we were conscious of the drawbacks of most symposium publications. All too often, individual authors present large quantities of data with little attempt at a synthesis, chapters are not well integrated with each other, and no attempt is made to provide a conceptual guide to the area under study. In this volume, the authors were discouraged from presenting large quantities of data, and were encouraged to speculate and develop theses with the aim of fostering interdisciplinary interactions. This volume, then, contains papers with a much higher ratio of speculation to data than is common in the literature. Our hope is that this will have the consequence of making the papers in each area accessible to specialists using many different approaches and that this, in turn, will foster collaborations, new concepts, and new hypotheses. We especially encourage students of predator-prey relationships to consider combining avenues of research that span traditional disciplinary boundaries.

Many individuals and organizations have generously contributed their time and efforts on behalf of the symposium, and we are grateful to them all. First, we thank the National Science Foundation for their grant (BSR83-20671), which made the symposium possible and provided funds for generating the camera-ready copy for the book on the University of Chicago's mainframe computer. We are also very appreciative of the efforts of all three professional societies (and their elected officers) that sponsored the symposium, and to the local meeting committee (chaired by Victor Hutchison) for their extensive assistance. Each paper was refereed by two individuals, and we greatly appreciate the time and effort that went into the reviews. Two reviewers also commented on the entire book, and their criticisms were very helpful to us. Other individuals whose assistance in preparing the book is greatly appreciated are: J. C. Briggs, J. Clark, C. Gilbert, J. Hives, J. Mambretti, and P. Wainwright.

1   Introduction

George V. Lauder and Martin E. Feder

1. Introduction

This is not a typical symposium volume. In most symposia, established investigators are brought together to summarize their research accomplishments and provide a detailed review and synthetic summary of their field. Such proceedings are usually characterized by the production of a symposium volume with a series of data-intensive papers that may include new findings, and the resulting symposium is often read only by close colleagues of the authors. In organizing this symposium, we insisted that invited speakers *not* present large amounts of data and *not* merely summarize the last ten years of their work. Our general aim was to provide a forum to consider critical concepts relating to predation from a variety of perspectives, and we asked each speaker to evaluate critically the contribution of his field to the study of predation. We specifically encouraged speculation, constructive criticism, and the discussion of novel and as yet undocumented ideas and hypotheses. This volume, then, is written expressly for students and researchers who wish to gain a broad appreciation of selected concepts and experimental approaches to the study of predation.

The symposium was organized to bring a variety of subdisciplines to bear on the problem of predator-prey relationships. As we envisioned the program, it would first consider functional and mechanistic approaches that are typically carried out in the laboratory, and then examine more naturalistic and evolutionary studies that are typically carried out in a field setting. The specific areas we chose to represent and the invited speakers were: organismal energetics (Albert

1

Bennett), sensory physiology (Gerhard Roth), functional morphology of feeding (Carl Gans), locomotion (Paul Webb), combined laboratory and field approaches centered about natural history (Raymond Huey), modelling (Earl Werner), empirical tests of models (Gene Helfman), ecology and behavior (John Endler and Gene Helfman), and quantitative genetics and population biology (Stevan Arnold). One paper, that of Harry Greene (Chapter 7), was not presented in the original symposium. We invited this paper subsequently as a result of this author's contributions to the discussion and the relevance of the topic to the issues addressed in the symposium. Also, Earl Werner, who did contribute a paper at the symposium, did not prepare a manuscript for publication.

We have endeavored to retain the flavor of the oral symposium presentations as a means of gaining a generality of discussion and of retaining speculation that might otherwise have been eliminated in a written paper. As you read this book, you should not expect large amounts of data nor should you expect comprehensive synthetic summaries of predator-prey relationships in lower vertebrates. Rather, we hope that this volume is successful in providing a stimulating general discussion, an introduction to perspectives on predator-prey relationships often omitted from general texts, and speculation that may point the way to new avenues of research and the formation of more general theories and concepts.

2. "Lower" Vertebrates and Predation

This symposium was held at the joint meeting in 1984 of the American Society of Ichthyologists and Herpetologists, The Herpetologists' League, and the Society for the Study of Amphibians and Reptiles. Although we restricted our contributors to those who focus on predator-prey relationships as they occur in lower vertebrates, we feel that the contributions have broad applicability to the general study of predator-prey relationships.

Research on the so-called "lower" vertebrates (fishes, amphibians, and reptiles) has provided an extraordinary wealth of information within the last decade on the relationships of predators and prey. Although theoretical issues and discussions certainly range beyond these taxonomic units, the "lower" vertebrates have been important in the formulation and testing of models of predation (Schoener, 1971). In addition, studies of the functional morphology and physiology of predation are well advanced in lower vertebrates, perhaps in part due to the amenability of fishes, amphibians, and reptiles for both laboratory and field experimental studies.

Studies of the physiology, functional morphology, ecology, and behavior of lower vertebrates have increasingly used modern analytical tools and experimental approaches to generate a comprehensive literature in many sub-specialties within the topic of predator-prey interactions. Progress has been especially rapid in specific areas such as optimal foraging (Emlen, 1966; Schoener, 1971; Hughes, 1980; Townsend and Calow, 1981), the functional morphology of predation (Gans, 1974; Lombard and Wake, 1977; Liem, 1980; Webb, 1982), the ecology of predation (e.g., Zaret and Rand, 1971; Stein, 1977), and the sensory physiology of prey escape response (e.g., Eaton et al., 1977; Eaton and Bombardieri, 1978). However, communication between investigators working in different areas has been inadequate. In addition, research on predation has become somewhat polarized with advances in functional morphology and physiology often remaining separate from similar progress in behavior and ecology. Some investigators simply ignore research in key areas. For example, Taylor's (1984) book on predation contains nothing on the mechanisms by which prey are captured or reference to the now well documented fact that an organism's sense organs and brain play a major role in determining which types of prey will be responded to as food (Chapter 4). The book focuses, instead, on ecological theory and predator-prey behaviors. Similarly, the theoretical models and "principles" of predation discussed by Holling (1959) contain little mention of the morphological and physiological factors that may constrain predatory behavior. Although Holling's "functional response" has become a widely cited and influential concept in predator-prey studies (Taylor, 1984; Hassell, 1978), it is still not clear how specific morphological and functional specializations for predation influence the validity of the models and results derived from the functional response paradigm.

Similarly, functional morphologists have commonly ignored important populational and ecological factors in their research programs. Variation between individuals in functional characteristics and the relationship of minor differences in function in the laboratory to actual performance in the field have not received the attention they deserve.

3. Goals

We have chosen the specific topics of the papers to reflect both areas of rapid progress and fields that possess data of wide potential applicability to general theories of predation. Thus the topics of sensory physiology, functional morphology, and organismal metabolism are included not only because of recent impressive discoveries, but also because emerging generalities within these fields are of considerable (and largely unnoticed) importance for analyses of the coevolution of predators and prey and for models of optimal foraging.

The major goals of this volume are: (1) to stress the importance of integrating information from various subdisciplines as general concepts and hypotheses emerge about the evolution of predators and prey; (2) to furnish concrete examples of the value of integrating physiological, morphological, ecological, and behavioral analyses within a system; and (3) to offer a critical analysis of the actual and potential contributions of several important research areas to the study of predation. We asked each of the participants to address in some form the following questions: "What is the utility of your approach to understanding predator-prey relationships?", "What are the strengths and weaknesses of your style of investigation into predator-prey relationships?", "What sorts of questions is your approach best suited to answer?", and "What are the major research problems that remain to be addressed using your techniques and conceptual framework?".

We do not anticipate that students in each of the specific research areas covered in this volume will be surprised as they read about their own area; indeed, they may find the discussion simplistic. But we do hope that they will find the concepts and discussions of other research areas stimulating and thought provoking. Above all, we hope to contribute to the conceptual evaluation of the benefits and problems inherent in alternative approaches to the study of predation.

References

Eaton, R. C., and R. A. Bombardieri. 1978. Behavioral functions of the Mauthner neuron. In *Neurobiology of the Mauthner cell*, ed. D. S. Faber and H. Korn, pp. 221-244. New York: Raven Press.

Eaton, R. C., R. A. Bombardieri, and D. L. Meyer. 1977. The Mauthner-initiated startle response in teleost fish. *J. Exp. Biol.* 66: 65-81.

Emlen, J. M. 1966. *Ecology: an evolutionary approach*. Menlo Park, Calif.: Addison-Wesley.

Gans, C. 1974. *Biomechanics: an approach to vertebrate biology*. Philadelphia: J. B. Lippincott.

Hassell, M. P. 1978. *The dynamics of arthropod predator-prey systems*. Princeton: Princeton Univ. Press.

Holling, C. S. 1959. Some characteristics of simple types of predation and parasitism. *Can. Ent.* 91: 385-398.

Hughes, R. N. 1980. Optimal foraging in the marine context. *Oceanogr. Mar. Biol. Ann. Rev.* 18: 423-481.

Liem, K. F. 1980. Adaptive significance of intra- and interspecific differences in the feeding repertoires of cichlid fishes. *Am. Zool.* 20: 295-314.

Lombard, R. E., and D. B. Wake. 1977. Tongue evolution in the lungless salamanders, family Plethodontidae. II. Function and evolutionary diversity. *J. Morph.* 158: 265-286.

Schoener, T. W. 1971. Theory of feeding strategies. *Ann. Rev. Ecol. Syst.* 11: 369-404.

Stein, R. A. 1977. Selective predation, optimal foraging, and the predator-prey interaction between fish and crayfish. *Ecology* 58: 1237-1253.

Taylor, R. J. 1984. *Predation*. London: Chapman and Hall.

Townsend, C. R., and P. Calow. 1981. *Physiological ecology: an evolutionary approach to resource use*. Sunderland, Mass.: Sinauer.

Webb, P. W. 1982. Locomotor patterns in the evolution of actinopterygian fishes. *Am. Zool.* 22: 329-342.

Zaret, T. M., and A. S. Rand. 1971. Competition in tropical stream fishes: support for the competitive exclusion principle. *Ecology* 52: 336-342.

2 Functional Morphology of Predator-Prey Relationships

Carl Gans

1. Framework

This volume deals with the relationships of predator and prey and is concerned with the contribution that various methods and approaches can make to our understanding of such systems. My task is to introduce the following chapters and deal with an approach known as functional morphology. It may be said disparagingly that before professors discuss anything, they become involved in the definition of terms. I would posit that this is not necessarily a bad thing, as long as means and end are not confused. Certainly, I have found that it is necessary to define purposes for myself before dealing with methods. Consequently, I propose to begin by placing my methods into the context of the whole organism and its interactions, in short by establishing what we mean by the relation of predator and prey.

In the present case, it should be obvious that each member of many predator-prey pairs is likely to be both a potential predator and a potential prey (Fig. 1). This hierarchical pattern is implied by, but may not be as rigid, as what ecologists refer to as trophic levels. It obviously implies a limit to the amount of "effort" that each organism may invest in improving its capacity either in success as a predator or in success in avoiding becoming prey.

In this framework, any predator must face a series of problems. Which resources are available, and where? What is their cost, in energy and in risk? How steady is their supply? Very important to any predator is that it must also avoid itself becoming a prey object in turn.

In its role of potential prey, the organism faces a different series of problems. What predators share its environment, and how frequently will these be encountered (per biotope)? Can they be avoided, can they be deterred, and which can be defended against in energy expended directly and in energy lost because the feeding or mating efficiency of the "prey" is reduced by the need for deterrence? What does each of these responses cost? How common and how constant will be the "attention" of a particular predator (i.e., is it a generalist or a specialist)? The prey organism must always balance the answers to such questions; it must balance the risk of predation against the effect that avoidance of predation will have on its ability to harvest resources.

Consideration of such cost/benefit relations should remind us that success has a price. As a particular predator becomes more successful (common or effective), it generates a greater risk for its prey; this makes any deterrent increasingly advantageous. Moreover, as a particular prey becomes more successful (and its biomass increases) other classes of predators will find it advantageous to pay more attention to it; thus there will be an increased advantage to overcoming its possible defenses. Evidently, predator-prey associations represent complicated systems involving cost/benefit relations for each prey type and each type of predator. In any biotope, the system is further complicated as it involves multiple species of predators simultaneously preying on multiple species of prey. It is also affected by density dependence and environmental uncertainty. Hence, the relation of costs and benefits must be solved for a state that is not only unsteady but intrinsically unpredictable.

This simple set of concepts should indicate why there must be many approaches to this kind of a multifaceted problem. Natural history data (Chapter 7) are required to indicate that predator-prey

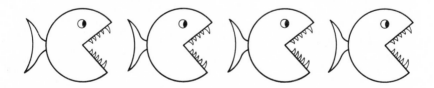

Fig. 1. Each predator may also serve as prey, suggesting that adaptation is unlikely to be only for either of these aspects independently.

interactions occur among particular species pairs. The sensory aspects of predator and prey recognition become important, both from the viewpoint of sensory physiology (Chapter 4) and of the behavioral (Chapter 9) and the defensive mechanisms (Chapter 8) that prey may utilize. The physical attack, capture and escape strategies are constrained by the mechanical and locomotor capacities of organisms, which may be studied from several viewpoints (Chapters 2, 3, and 5). Aspects of energetics may also be significant (Chapter 5). Tests of the reality of our preliminary conclusions about associations of structures and roles demand comparative studies and the manipulation of systems (Chapter 6) as well as information about the heritability of the underlying traits (Chapter 10).

2. Analysis of the Phenotype

Biologists generally start the analysis of a particular case by an implicit or explicit examination of the phenotype or structure of the organism; they can observe more or less easily that this permits certain actions to occur. Limbs can swing. Hair may wave. Rhodopsin can bleach. Scales can shine. Neurons can pass signals from bouton to postsynaptic region. All such actions are referred to as functions in common parlance. However, some of them might be irrelevant to the survival of the organism. Others are of use to the animal and consequently are acted on by natural selection. They presumably are the basis of adaptation. Such adaptive aspects of the phenotype are referred to as the roles (Bock and von Wahlert, 1965; Gans, 1985). Their recognition and demonstration is often difficult.

Analysis of any specific case of predator-prey relationships obviously requires characterization of the way in which a potential predator attempts to harvest prey and a potential prey attempts to circumvent this action. Involved is a combination of structural mechanisms and of behaviors (and physiologies) which they may support. I have already noted that the actual situation is often complicated by the interaction of multiple individuals and species, both of the predators and the prey, and by the diversity of the physical and biotic environments against which these plays proceed. Nevertheless, we must begin by studying a phenotype and the functions and roles associated with it.

In the following analyses the species' phenotype is referred to as if it were a constant unit, the function and role of which are to be studied. However, it is recognized that ontogenetic and sexual differences occur. Moreover, the phenotypes and genotypes of any biological population vary and the variants are likely to differ in their capacity to support roles. Selection among variants is potentially a critical component in the development of adaptation (Chapter 10).

It is important for the present discussion that such systems should not be analyzed in an overly abstract sense. In any situation, we are not modeling a theoretically perfect insect catcher, which was designed "to operate among a particular assemblage of rocks or tree trunks". Rather, the practical case is likely to demand that a population of iguanids be able to maintain a genetically stable number in some valley of the Chihuahuan Desert by harvesting fluctuating populations of flowers and insects in the presence of competitors, predators and climatic changes.

In studying such systems, we should keep foremost in mind the difference between function and role. Does the behavior allowed by a structure increase the fitness of the particular individual, or is the behavior merely permitted by (and represents an incidental effect of) a structure formed and maintained for quite independent causes? The cause may be selection for separate and different functions or constraints due to the genetic or developmental pattern, i.e. due to the way the phenotype forms during ontogeny.

Analysis requires the nested series of approaches discussed below. It is unlikely we could find species pairs that differ in only a single set of structures and a single associated set of functions. Multiple differences in structures and functions likely characterize most species. Consequently, multiple sets of comparisons are required using multiple species to resolve multifactorial problems; the approach is analogous to that of mathematicians who must use as many equations as they have unknowns. It is best to analyze such patterns from a phylogenetic perspective as this will tell us how often a particular characteristic arose. The methods of comparative biology offer the best tools for approaching this question (Gans, 1985).

3. The Place of Functional Morphology

Study of a predator-prey system normally starts with the observation that two or more species feed upon each other (see Chapter 7). The obvious first steps of any analysis are (1) the characterization of their phenotypes, (2) detailing of particular behaviors and physiological aspects observed, (3) a check of whether and how (1) facilitates or permits (2), and (4) the establishment how close the match is. The third and particularly the fourth of these steps benefit greatly from ecological and interspecies comparisons.

Functional morphology provides techniques for evaluating which aspects of the phenotype allow the observed behaviors and physiological functions, and how these phenotypic aspects are involved in these functions. This involves characterization of the phenotype and of the behavior. Once it has been completed, it is possible to test the closeness of the match between structure and function. For instance, the observation that several variant dentitional phenotypes (structures) characterize a group of species all of which share a general method (function) of behavioral/physiological food reduction suggests matching of structure and function. However, if each behavioral variant is reflected in a morphological one, the matching obviously would be closer, lending strength to the hypothesis that each function reflects a biological role. Hence, studies in functional morphology provide a basis for approaching issues of adaptation.

Also, functional morphology provides the basis for a consideration of the transition of species through time and allows evaluation of the differences observed among organisms belonging to distinct higher taxonomic categories and to varied ecological associations. It approaches the question of whether particular interspecific morphological differences may be associated with other differences, for instance different behavioral and physiological functions. Functional morphology also allows us to collect evidence on the mechanically possible, which is important because not all structurally intermediate states prove to be functional. For instance, reduction (or increase) of the absolute size of an animal by geometric scaling down may be disadvantageous both in terms of the mechanical and the developmental programs; hence, there may be selection for change in proportions (expressed as different and size-dependent coefficients of allometry).

It is, of course, possible to argue that this kind of analysis cuts inappropriately across the classical subject areas in which organisms tend to be studied. Thus, description of phenotype may be characterized as morphology, as development and as morphological genetics; detailing of animal actions may be studied as ecology and behavior; and functional description may reflect physiology and the newer area of biomechanics. However, I would posit that such subject divisions are artificial. To quote a comment Dan Janzen once made to me, "they are artifacts of the way we do biology". If the aim is to understand organisms, we must transcend this level.

Functional morphology then is a way of matching phenotypes to tasks. It produces information about which physical-chemical principles may be involved in allowing phenotypes to perform functions. The functions of organisms are not subdivided into neat categories; animals are likely to utilize physical, chemical and biological principles in an unpredictable sequence. The likelihood of a correct decision regarding structure-function relations hence increases with the investigator's access to the combined principles of these disciplines. Territorial subdivision of the field does not foster insight. Hence, this symposium.

Functional morphology thus should involve a holistic approach to the organism, and holism is one of its great potential advantages. Matching of phenotypes to tasks has traditionally been begun by subdividing the organism into gross mechanical linkage units, studying them as well as the muscles that move them. We have seen independent analyses of teeth as holding devices and teeth as venom injecting devices. However, combined analyses within the framework of the ecological demands upon the organism have taken us from narrow "either or" statements to more robust "relative cost-benefit" solutions. More recently, studies of the hydrodynamics of propulsion, for instance the aspects described by Dr. Webb (Chapter 3), allow us to evaluate the relative benefits to fishes of a fast start. Furthermore, functional morphology is able to proceed on other levels as well. It can include analyses on the tissue level (oxygen extraction mechanisms present an obvious category), on the sub-cellular level (salt glands permit feeding on salt-loaded food and predation in salty environments), and on the biochemical level (matching the structure of venoms to particular prey types).

All of these levels are important, because closeness of matching between role and structure remains the prime problem, and the one for which the tools of functional morphology are the most useful. Decision about the function-role "alternatives" requires detailed description of structure, environment and function. Only this data base permits us to frame questions about possible roles and the degree to which these may be matched to the phenotypes.

In this sense, the discipline of functional morphology utilizes a broad array of techniques, many of which were developed as part of other fields and are here combined to achieve answers. For example, structure is characterized by x-raying organisms and by cutting or sectioning them to confirm details of internal topography. One identifies components by staining them, by chemical assay, by dissolving away the "unnecessary" and conserving the "significant"; the "unnecessary" may be the soft tissue in the study of bone or connective tissue in the study of muscle.

Function is approached by observation and measurement. Movement, force, electrical events, and chemical changes are recorded by cinematography, cinefluoroscopy, force plate and strain gauge recording, and electromyography, measurement of fluid flow and metabolic rate. In short, many physiological techniques are likely to be usable, the limitation being that too detailed initial analysis may conflict with the need to look broadly enough to understand more than a restricted part of the system.

4. An Example of Analysis: Muscle Arrangement in Shingle-Back Skinks

4.1 Problem

It may be useful to consider the breadth and depth of functional analysis by using examples from analyses of the muscular system. Muscles can be and have been compared by describing their origin and insertion, their shape and weight. Considerations of fiber architecture and histochemistry provide more information. Recognition that muscles are subdivided into task groups awakens us to the possibility that a particular muscle may contain two or more functionally different units. Thus, in some animals, the muscular system is subdivided into separate groupings, some with muscles providing support, and others producing either fast or slow propulsive forces. In other animals, a single muscle carries out all of these functions. Detailed information about the way muscles are organized is obviously needed for developing sound structure-function-role correlations.

I should like to mention here a stage in our recent work on mastication in an Australian lizard, the shingle-back skink (*Trachydosaurus rugosus*), as an example of possible levels of analysis (Gans et al., 1985). During my stay in Western Australia, I became interested in the feeding behavior of this large lizard because *Trachydosaurus* was locally common and because the shape and width of its head reminded me of that of the tuatara, *Sphenodon*. Our earlier study of *Sphenodon* had disclosed a peculiar food cutting mechanism (Gorniak et al., 1982). The dentary tooth row of *Sphenodon* closes into a gap between the rows of maxillary and palatine teeth. The mandible then slides forward, thus cutting the prey by a shearing action.

The slow and heavily armored *Trachydosaurus* marches deliberately up to potential prey, biting and crushing it. The behavioral literature on *Trachydosaurus* documents its omnivory. Food items mentioned include flowers, tomatoes, beetles, snails and carrion. Superficial examination of specimens of *Trachydosaurus* disclosed that their dentitional pattern is distinct from that of the tuatara (like lizards generally they lack palatine teeth) and that the skinks seemed to differ further from tuataras in not sliding their mandible forward after closing the mouth.

4.2 Initial Analysis

(1) Characterization of Structure. Perhaps the most striking anatomical feature of the mandibular system of *Trachydosaurus* was the substantial and relatively complex adductor mass. This muscle mass differed from that of other lizards in the variable length and angle of placement of its fibers. Examination of the skeleton of *Trachydosaurus* suggested that the mandible could make only simple open-close rotational movements. If this were so, why were the muscles so large and complex?

(2) Characterization of Function. Observations of the behavior of *Trachydosaurus* indicated that as predators they could crush large and resistant objects (such as snails and bones) and that as prey they would display with the mouth wide open to deliver a forceful bite, even at large objects. The force of the bite (at a wide-open position) thus might be a possibly significant aspect. Thus, one possible explanation for the structural pattern of the musculature was that it reflected the necessity for having a large number of fibers arranged to act "in parallel". This hypothesis might explain the bulk; did it also explain the complexity?

Electromyography (EMG) showed that the major portion of the adductor mass was active in synchrony. Furthermore, the EMG signals from the adductors did not increase substantially whenever the lizards were crushing hard prey even though one assumes that this required greater force production. Also, if very large snails were being crushed, the head pointed downward; the symphysis briefly but regularly then contacted the ground. This jaw contact suggested that the ventral neck muscles might be involved in crushing action and EMG confirmed that this was the case. A more important observation was that action of these muscles was synchronized with that of the adductors only when the symphysis was in contact. Whenever the head was off the ground the muscles fired in opposition ruling out incidental coactivation. The

EMG's of the adductor mass did not show pulsatile activation whenever large and hard snails were crushed.

(3) The Nature and Closeness of the Structure-Function Association. Initially vexing was the observation that the EMG signals from the adductors did not show amplitude changes but pulsations (at frequencies near 9 Hz). The possible significance of the pulsatile activity pattern (Fig. 2) was assayed by synthesizing it and feeding it back to the adductor muscles through stimulating electrodes. Pulsation at the observed frequency produced a stepwise increase of force. The level attained was several times that observed in simple twitch activation. This confirms that for snail crushing the muscles are activated tetanically. Moreover, adjustment of the number of pulses applied to the prey might allow a lizard to match the muscular force to the breaking strength of the prey.

A series of prey crushing experiments further showed that the snails which represent a major food item could only be broken by the bilateral application of substantial forces. The force levels required for crushing apparently exceed those which *Trachydosaurus* can generate with the muscles of one side of the jaws, so that the entire jaw musculature is involved in reducing this prey. Such observations, coupled with the apparent recruitment of the ventral neck muscles, suggests that the large size of the musculature can be explained by the need for maximal force generation. This explanation would seem to have a better explanatory value than alternate hypotheses, such as that the large muscular masses are only advantageous as a filler to maintain the cephalic contour of the head or that the mass is established due to some allometric drive.

The relatively simple movement of the mandible has been confirmed by cinematography and cinefluoroscopy. The problem of why the adductor mass is so complex remains. One possible explanation for this complexity is that different portions of the muscle act separately. However, EMG indicates that all the muscles and fascicles act essentially simultaneously. There is no temporal subdivision that might reflect a basis for separation, such as might be expected if complexity reflected functional separation. Of course this assumes that the fiber properties are equivalent, and that motor units are activated equivalently; we know these assumption are unlikely to be true.

Fig. 2 (facing page). Electromyograms derived from six mandibular muscles of a specimen of *Trachydosaurus rugosus* crushing a snail. Note the pulsatile activation during the initial crushing bite.

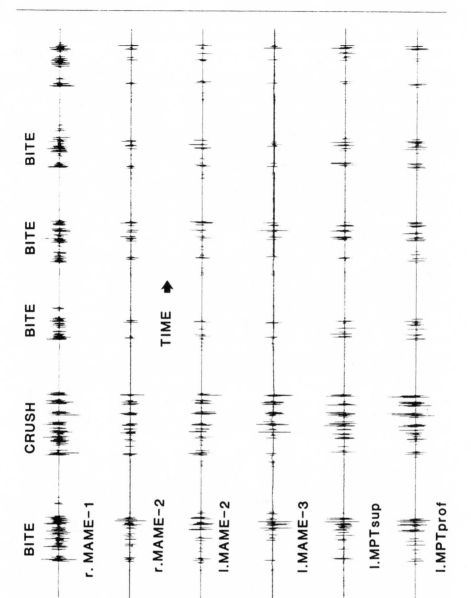

Another possible explanation suggests that the muscle architecture is invariant among species, so that the pattern is group specific rather than being subject to selection in each species. However, previous anatomical analyses (Gomes, 1974) characterize significant differences among the species of *Tiliqua*, reducing the likelihood of this explanation.

Falsification of possible alternatives leaves the placement of multiple fibers in parallel as the only credible explanation for the complexity. The question then becomes: Is it possible to match observed fiber length and fiber angle to any other factor? Packing of fibers is an obvious possibility. The insights to which this led are detailed in the next section.

4.3 Packing and the Length-Tension Relation

A stimulated muscle fiber exerts a force on its ends, the magnitude of which is a function of the length-tension relation of its component sarcomeres (contractile units). Any mandibular adductor fiber produces a moment, the magnitude of which is dependent on the normal distance from the site of rotation to the line of action. Depending on the angle and position of insertion of the fiber, the moment arm may increase or decrease with opening. Shifting the position of the resting or plateau length of the sarcomere relative to the degree of opening of the jaws allows the muscle to generate maximum moment anywhere between full open and full close. Hence, the moment generated by muscular contraction may be greatest near occlusion (as in man) or near the fully open (threat or snail crushing) position as measured in *Trachydosaurus*.

For initial analysis, it is assumed that the individual sarcomeres stacked end to end along the fibrils of a muscle are physiologically equivalent and will at any instant all occupy equivalent positions on the length-tension curve. It is then easy to demonstrate that parallel placement of the fibers of the mandibular adductor muscle between the planes of insertion and origin would only result in the formation of a simple wedge (Fig. 3). The longer fibers within such a wedge would have more sarcomeres in series, the number being in proportion to their length (in turn proportional to the distance of their point of insertion from the site of rotation). Expansion of the muscle, by utilization of the volume dorsal and posterior to this simple wedge for fiber placement, requires tendons and aponeuroses and likely makes the muscle pinnate. Also, the shift of origin and insertion away from the surfaces defining the wedge would change the moment arm of the additional fibers unless their angle (relative to the mandible) changes as well. Consequently, it is expected that in a muscle formed of equivalent sarcomeres, the insertion site of each fiber will determine the direction of the moment arm and presumably the angle at which the fibers will be aligned. The length of moment arm of each fiber will continue to determine its length and hence the number of sarcomeres in series.

One unexpected consequence of this is that for such a muscle of equivalent sarcomere arrangement, each sarcomere will contribute equivalently to the overall moment that the jaw will impose on the prey. These considerations provided a set of predictions that could be compared with the architecture observed in the main mass of adductor

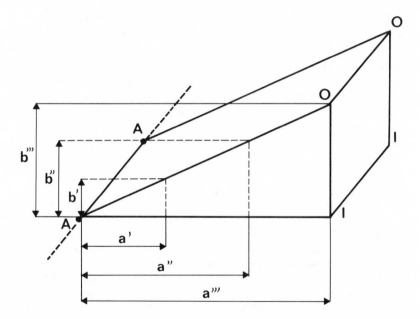

Fig. 3. Wedge of fibers tending to approximate one surface (AAOO) to another (AAII) by rotational action about axis A-A. The number of sarcomeres of each fiber, placed in parallel in such a wedge, is likely to be proportional to the length of the fiber (b', b", b''') and hence to the moment arm (a', a", a'''). Each sarcomere, in such a wedge of fibers placed in parallel, is likely to produce equal moment for equivalent shortening and to induce equal shortening for equivalent moment. Hence, the common assumption that muscles placed more distally from a joint are more advantageous is not likely to be true for most cases.

muscles. Once fiber placement was corrected for distance from joint and angle relative to the surface of insertion, it appeared that the fiber length was highly correlated within most muscles. Moreover, the main adductors provide maximal force at a close to wide open position.

However, the model does not explain the arrangement of all muscles. Some sets of medially and laterally displaced fibers tend to be very short relative to the length of the entire muscle which inserts at positions at which it generates a low moment arm. Such low moment arms may result either from insertion close to the site of rotation or more distantly, but at a shallow angle. As the linear muscular shortening required to effect unit angular rotation is proportional to the moment arm, the sarcomeres of these short fibers may still remain in an advantageous position of the length-tension curve. Obviously a given mass of muscle generates more force if arranged as parallel short fibers than as an equivalent muscle mass of longer ones.

The analysis allows quantification of the effectiveness of different patterns of muscle placement. Geometry establishes how many fibers (and of what length) may be packed in parallel between two surfaces (of origin and insertion) at any angle to the line of action of the muscle as a whole. This permits the simple calculation of whether the loss of force due to pinnation of the fibers is compensated for by the increase in their total number. The test can be applied to the Mm. pterygoideus and adductor mandibulae externus I. In the latter muscle, the shallow insertion reduces the effective moment arm of the entire muscle by a

Fig. 4 (facing page). The lines of action of muscles placed shallowly (i.e. at angles much less than 90° to the action line of the rotating element) incorporate adversely placed vector resultants that act along the length of the rotating element and may not induce rotational components. Such resultants do induce compressional forces that may generate inappropriate loading of the joint, but may be balanced by the resultants of other forces. As the magnitude of the resultants changes with degree of opening, the calculation for the force pattern must consider the time (determined by electromyography) at which the muscles are active as well as the resultants of all force vectors induced at that time. The top view shows the skull of a lizard (*Trachydosaurus rugosus*) with the mouth closed and the bottom view of the same specimen with the mouth open. The force vectors (all assumed to be equal) show the lines of action of the major shallowly inserted muscles (Mm. adductor medialis externus - MAME-I and pterygoideus) resolved along the lines of closing action and along the length of the mandible (a vector that might lead to mandibular dislocation). The bottom picture also shows the vector of the major opening muscle.

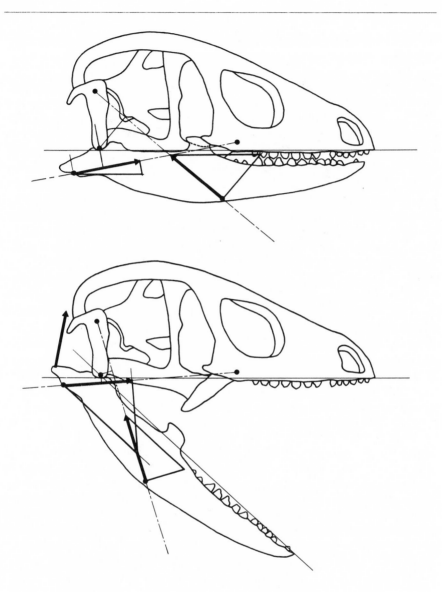

factor of 3 or 4, and the pinnation further lowers the contribution of each fiber by some 30%. However, the number of fibers that can be placed between origin and insertion increases at least 10-fold. Thus this arrangement at least doubles the potential force of closure and does so with a muscle placed in a less than prime position.

The packing pattern is not the only factor that needs to be considered to understand the functional basis of this muscle architecture. The very shallow insertion of powerful multipinnate muscles not only induces major forces in the closing direction but also longitudinal ones along the mandible (Fig. 4). These might tend to disarticulate condyle from fossa, particularly whenever the animal bites forcefully with the jaws gaping widely. In *Trachydosaurus*, the lines of action of two such muscles with a shallow articulation cross each other so that their horizontal vectors have the potential for balancing and unloading the jaw joint. This reptilian arrangement also provides us with an analog to the crossed lines of action of the mammalian masseter and temporalis muscles, which then represents an analogous condition that developed anew after the change of jaw joint (Crompton, 1963).

4.4 The Next Steps

Functional morphology has, in this case, provided several functional explanations of the way that the mandibular musculature of one species of lizard is arranged and used in prey handling and other forceful bites. It certainly disclosed that the reduction mechanism is quite distinct from than seen in the tuatara; *Trachydosaurus* cannot cut and can only reduce food into pieces by a forceful bite, coupled with a lateral shake. Although this level of analysis clearly does not establish the role of structures, it provides the basis for further comparative and ecological studies in which the role may be tested. In combination with these, it may disclose the utility of parallel studies of the functional morphology of related species and those with analogous mechanisms. Are the slow response patterns and heavy armor always associated with a strong defensive bite?

Questions requiring ecological study include the importance of hard-shelled prey in the overall dietary repertoire of this lizard in comparison to other species that show a less forceful mechanism. Do the wide gape and crushing capacity actually have an effect on the range of foods consumed? Indeed, is the force amplification related to prey at all or does it reflect mainly a need to bite in defence? Is there sexual dimorphism in head shape suggesting male combat? The kinds of field observations and tests described in Chapters 6 and 7 become important at this stage, because the present questions require information about other functions in which the phenotype is involved. How often is each phenotypic aspect utilized and how significant is each utilization to the life of the organism?

The comparative approach then becomes critical. Are there related species? Do these disclose morphological, behavioral and ecological equivalence and to what extent are the differences among them correlated? Are there nonrelated species that show similar morphologies, behaviors and ecologies? As stated earlier, the establishment of possible correlations demands that the structure-function match be performed simultaneously for multiple species.

The identification of role remains the key outstanding confounding factor. It is possible to demonstrate that a phenotype permits a function and even to establish how the phenotype permits it, but difficult to demonstrate that the phenotype that allows this function is not shaped or maintained by a different role. In order to answer such questions one may begin to ask about the other functions in which the phenotype is involved, about their significance, how often it is utilized for them and how advantageous or how important is each utilization to the life of the organism.

5. Conclusion

In short, functional analysis based upon a single species in a laboratory setting cannot by itself identify the roles. Study of the functional morphology of additional species, as well as ecological observations and physiological measurements (preferably under field conditions and in a comparative setting), are required to establish this. Such approaches may provide distinct and supplementary information. Many are discussed individually in subsequent chapters, although it is often desirable to combine several such approaches to understand the pattern seen in a single species.

Much of the phenotype that participates in a component function is likely to be matched to it, i.e. the phenotype will be permissive. It is furthermore obvious that structures will participate in multiple roles and the resolution of these roles may be difficult. Indeed, the influences of the multiple functions which we see for different aspects of the phenotype are likely to be additive and conflicting, and more likely to be subtle in their effect on the adaptiveness of the individual. Of course, those aspects of the phenotype subject to seemingly overwhelming environmental influences are more easy to test. For example, the unforgiving demands of deserts or of aerial flight may allow only a limited number of solutions and these are likely to be major in their effect on physiology and geometry. Consequently, their influence is easier to recognize in small samples, than might be the influence of more subtle selective effects.

The balance among selective effects furthermore changes throughout the life history of the individual and differs among sexes. Not all functions may be significant to a particular individual. Selection on rare aspects during the life history of some potentially reproductive individuals may still be significant.

I do hope that I have been able to demonstrate that analysis of the adaptations involved in predator-prey systems had best start by characterizing the phenotype. A detailed analysis of its possible functions remains important. Hence, it seems appropriate to turn about a saying of the late D. Dwight Davis and to note that functional morphology is the handmaiden of *all* of the branches of evolutionary study. Morphology lacking a comparative and functional perspective is sterile. Function lacking a basis of descriptive morphology is likely to be inadequate.

Acknowledgments

I thank the editors and an anonymous reviewer, as well as S. Arnold, D. Carrier, H. W. Greene, R. B. Huey, P. Pridmore and P. Webb for multiple comments on versions of this manuscript. This presentation was supported by NSF Grant DEB 81-21229.

References

Bock, W. J., and G. van Wahlert. 1965. Adaptation and the form-function complex. *Evolution* 19: 269-299.

Crompton, A. W. 1963. Evolution of the mammalian jaw. *Evolution* 17: 431-439.

Gans, C. 1985. Differences and similarities: Comparative methods in mastication. *Am. Zool.* 25: 291-301.

Gans, C., F. de Vree, and D. Carrier. 1985. Usage pattern of the complex masticatory muscles in the shingleback lizard, *Trachydosaurus rugosus*: A model for muscle placement. *Am. J. Anat.* 173: 219-240.

Gomes, N. M. B. 1974. Anatomie comparee de la musculature trigeminal des lacertiliens. *Mem. Mus. Natl. Hist. Nat. (N.S.), A (Zool.)* 90: 1-107.

Gorniak, G. C., H. I. Rosenberg, and C. Gans. 1982. Mastication in the tuatara, *Sphenodon punctatus* (Reptilia: Rhynchocephalia): Structure and activity of the motor system. *J. Morph.* 171: 321-353.

3   Locomotion and Predator-Prey Relationships

Paul W. Webb

1. Synopsis

Predator-prey interactions invariably involve locomotion. Locomotor capabilities are determined by physical laws, which in turn limit locomotor behavior. Locomotor capabilities and performance can be examined in detail in the laboratory and the results can be used to predict optimal behaviors that should maximize the chances of a predator catching prey or of prey escaping an attack. Mechanical principles underlying locomotion can also be used to recognize situations in which such predicted optimum behavior cannot occur (inequalities) and to interpret observed locomotor behavior during predation events. This locomotion-based biomechanical approach to the study of predation is illustrated for fish piscivores.

The approach has some drawbacks. Test encounters are usually staged in artificial arenas and therefore should be complemented by (and complement) field studies. In addition, no single approach such as biomechanics is adequate to fully understand predator and prey behavior, necessitating use of methods from other disciplines in tandem.

Experimental studies of locomotion of predators and prey may be used to study interactions between animals from widely different taxa and changes during ontogeny. At this time, there is a major need for studies on motor activities actually used in predator-prey interactions.

2. Introduction

Interactions between predators and their prey are examples of interception and avoidance situations known as differential games. The participants in such games have opposite goals; a predator succeeds by catching prey while successful prey escape. Predators and prey can "play" the interception/avoidance game in many ways with the result that predation cycles are diverse and complex. However, locomotion plays a role in every situation. Stationary camouflaged prey are captured by wide-ranging, searching predators. Animals moving in search of food or in other activities are caught by ambushing or searching predators. Prey often flee and are run down.

Locomotor performance is constrained by well known physical laws. These laws can be used as a framework for the experimental examination of locomotor capacities. Fortunately, animal locomotion has been widely studied for its own sake and consequently there are a large number of approaches and techniques that can be used to understand predator-prey relationships. The physical properties of locomotor structures can be measured using a variety of techniques recording strains under static and dynamic stresses (see Wainwright et al., 1976). Animal movements and energy fluxes can be recorded at a variety of activity levels, and the limits of performance measured (Chapter 5). Mechanical analyses based on records of animal movements on movie film or video tape are widely used to evaluate and describe locomotor capabilities and limits. The result of such research is a picture of what animals can do. Nonetheless, because these techniques have been developed to understand locomotion rather than animal performance during predator-prey encounters, there are inevitable problems in application and interpolation, as discussed below.

Knowledge of an animal's locomotor capability can be used at several levels to understand predator-prey interactions. The first (and more simple) level is a sophisticated description of predator and prey behavior. For example, Elliott et al. (1977) showed that relative acceleration capabilities were critical to the success of lions preying on various animals. Vinyard (1982) evaluated the effectiveness and energy cost of two mechanically different motor behaviors used by Sacramento perch to catch evasive and non-evasive prey. Webb (1984a) used biomechanical principles to correlate locomotor form with foraging patterns of fish.

Description of locomotion during predation events typically leads to a second level of study in which research seeks to predict locomotor behaviors that should maximize the success of the predator and the prey. In general, a predator should attempt to minimize the duration of an interaction and prey should maximize that duration (Isaacs, 1975). Because prey often can reach cover in seconds, this means that a predator should try to take prey before the prey reaches cover. By contrast, the longer the prey can extend an interaction, the greater are its chances of reaching cover to "leave the game." Given the simplicity of these objectives for a predator and its prey, it may be relatively easy to relate mechanical constraints on locomotion to performance levels and behavior of participants, and hence to the outcome of interactions. This leads in turn to prediction of optimal behaviors, such as speeds and turning radii (maneuverability), that favor prey capture or prey escape (Howland, 1974; Webb, 1976; Weihs and Webb, 1984). In addition, when optimal behaviors can be predicted, special situations can be identified where such behaviors do not appear possible. Heinrich (1983) calls these cases "inequalities", and since they represent initially unexpected situations, he considers them particularly important for advancing theory.

Predictions based on biomechanical locomotor studies often lead to experimental tests of those predictions. These typically involve staged encounters between predators and prey in arenas appropriate for the hypothesis being tested. Arenas range from artificial enclosures in the laboratory through fenced-off areas in the field (Major, 1978) to observations of free-ranging animals (Elliott et al., 1977). However, biomechanical studies describe only what animals can do, and not necessarily what they actually do, so that experiments are rarely fully conclusive. Commonly, additional theory is used to reconcile expectations from laboratory studies with realities of behavior so that predictive and explanatory approaches often appear in tandem.

Thus locomotion is common in predation events and is easily studied in the laboratory. Research on the functional morphology of locomotion provides concepts and data that can be used to predict behaviors, identify inequalities, aid in experimental design, facilitate evaluation of observed behavior, and thereby lead to better understanding of predation behavior. My studies of predation by piscivorous fish illustrate these points.

3. Predator-Prey Interactions

3.1 Predicting Behavior

To analyze predator-prey interactions, the predation cycle can be divided into several phases (Fig. 1). Some phases are especially valuable starting points for predicting behavior from biomechanical analyses of locomotion. For two reasons, the most logical starting point is the strike phase. First, the strike is a pivotal point because predation cycles must eventually pass through this phase. Second, performance is most likely to be maximized during the strike. Prey in particular would be expected to use maximum accelerations and speeds to reach cover and "leave the game", thereby avoiding capture. Predators should move at high acceleration rates and speeds to try and catch the prey before the prey can reach cover. Thus, performance in the strike phase would be expected to approach the physical limits for locomotor systems, which in turn should place restrictions on behavioral options.

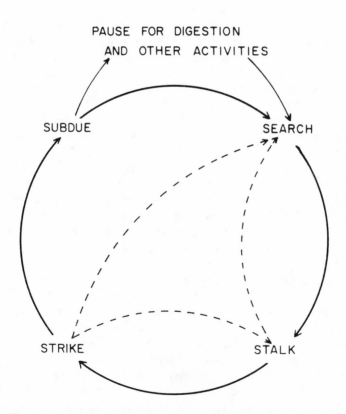

Fig. 1. The phases of a generalized predation cycle. Solid lines show transitions from phase to phase for successful predation events. Dotted lines show possible predator behaviors following successful prey avoidance.

Attempts to predict predator behavior for the strike must start with prey escape movements. Although a predator initiates a strike, subsequent events are determined by prey responses. A predator should assume prey will attempt to escape and should utilize approaches that are most likely to counter the prey's response. Hence predator tactics should be based on expected prey avoidance behavior.

Predators that are successful in approaching prey are likely to cause the prey to startle and flee. In fish, this startle response is usually initiated by one of two Mauthner cells. The Mauthner cell causes the whole myotomal muscle on one side of the body to contract, thus bending the body into a C- or J-shape. The body shape at the end of this first stage of the motor response gives the acceleration pattern the name *C-start*. Stage-1 is followed by a second stage contraction of the whole myotome on the other side of the body, thus bending the body into a second C-shape in the opposite direction (Fig. 2). These two stages are followed by continued swimming or a glide. Since the first two stages of the startle response involve successive contractions of the whole myotome on each side of the body, the response is a maximum acceleration. During this period, lasting 60 to 100 msec, fish can achieve peak acceleration rates of 4 to 5 g, and reach speeds in excess of 1.5 m/sec.

The body movements generate reaction forces normal to each point along the body length. Each such point is called a propulsive segment or element. During the first stage, water is mainly pushed sideways so that lateral forces are large. The resulting moment causes the center of mass to rotate and turn the body as it begins to move forward. In stage-2, propulsive elements face more caudally relative to the head so that turning moments are smaller and a larger portion of the normal force generates forward thrust. However, because of the initial rotation of the head, the trajectory of the center of mass remains at an angle to the original axis of the fish. Thus the startle response is both an acceleration and a turn. Furthermore, the angle through which the path of the center of mass turns is proportional to acceleration rate (Weihs, 1972, 1973). Since a startle response is a maximum acceleration, it follows that the angle of the turn of the center of mass is also maximum and relatively invariant.

Therefore, the all-or-none nature of the startle response results in initially stereotyped propulsive movements (Eaton et al., 1977), in predictable forces on the body and in predictable trajectories. The latter should be exploited by a predator and hence knowledge of the forces acting on the prey and of the trajectories of various parts of the body provides the basis for predicting optimum predator behaviors (Fig. 2).

Such behaviors concern the predator's choices of a target on the prey body, its strike trajectories and strike performance level.

a) *Target.* The center of mass of the prey initially moves less than the rest of the body and has the most predictable path (Eaton et al., 1977; Webb, 1978). A predator striking at this point on the prey's body as a target is therefore most likely to catch the prey.

b) *Strike trajectory.* A predator should choose a strike trajectory aligned with the anticipated escape trajectory of the target. The prey s center of mass starts to move along a path at about 80-90° to the initial prey axis. Turns can be started by the head in stage-2 of the response, but the path of the center of mass lags behind the head, maintaining

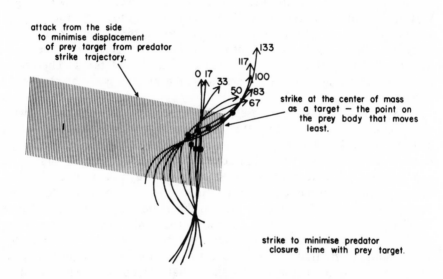

Fig. 2. Tracings of the center line of a fish during C-start acceleration in a startle response. The center of mass is shown by the solid circle. Time between successive tracings is shown in msec. Stage-1 lasts from 0 to 57 msec, and stage-2 from 67 to 133 msec. Since the startle response involves successive contraction of the whole myotome on each side of the body, escape movements are initially stereotyped. This leads to the recognition of three tactical components for predator strike behavior. These have been recorded for *Esox* attacking fathead minnow prey. This predator strikes at the center of mass as a target, using maximum acceleration. The average observed strike trajectory is shown by the shaded area, the width of the shading corresponding to the width of the mouth. The predator strikes from the side. Thus, the escape trajectory of the prey target is most likely to coincide with the strike path. the initial target trajectory

for longer than would be expected from the gross movements of the body (Weihs, 1972; Fig. 2). Therefore, a predator should strike prey from the side so that strike and escape trajectories are likely to coincide, maximizing the time prey are on the strike path.

c) *Performance level.* A predator should strike as rapidly as possible to minimize the duration of an interaction, minimizing the time to reach the prey and the opportunity for the target to leave the strike path. The most logical way to achieve such minimum closure times between the predator and prey is for the predator to accelerate maximally and directly towards the prey. However, this is not possible using the C-start accelerations seen in the prey startle response because the head turns to one side. Predators usually prepare to strike by aligning the body with the prey. If the predator were to strike using a C-start it would first move away from, not directly towards, the prey. A second fast-start pattern, called an *S-start*, avoids the problem. In S-starts, the body is bent into S-shapes and turning forces are distributed along the body, eliminating uncontrolled turns. Therefore, piscivorous fish should use S-starts to strike at their prey at maximum acceleration rates.

Experimental tests of these predictions have been performed with an esocid, the tiger musky (Webb, 1980; Webb and Skadsen, 1980). This is a hybrid of northern pike (*Esox lucius*) and the muskellunge (*Esox masquinongy*), both piscivorous ambushing predators. The experiments showed that *Esox* struck at prey as predicted from biomechanical arguments, especially in terms of expected targets and performance level. *Esox* struck prey in the region of the center of mass and used rates of acceleration that were the same as forced maximum levels. Strike angles were more variable, but in general the predator stuck along paths that would reduce the possible effectiveness of prey escape trajectories. These results show that a locomotor biomechanics approach can predict behavior.

3.2 Identifying Inequalities

The results from experiments on *Esox* also lead to new problems because many fish, especially acanthopterygians, are too short or stiff to bend into large amplitude S-postures that are required for S-starts. Therefore, such piscivores cannot use the predicted optimum strike behavior found using *Esox*. What behaviors are expected for short-bodied piscivores?

Solutions to this problem require comparative experiments. Four piscivores, tiger musky, rainbow trout (*Salmo gairdneri*), smallmouth bass (*Micropterus dolomieui*) and rock bass (*Ambloplites rupestris*), were observed attacking fathead minnow prey (Webb, 1984b). These predators have various body forms and motor capabilities for acceleration (tiger musky), sustained swimming (trout) and maneuverability (rock bass), with smallmouth bass being intermediate between the latter two species. The predators cover a range of locomotor abilities illustrative of bony fishes, although extremes (e.g., tuna) were excluded because of the difficulties of working with them.

The general expectations were that S-start acceleration strikes typical of esocids would be replaced by high-speed chases and rapid, small radius turns powered by the body and fins. These would maximize swimming velocity and centripetal acceleration rates, respectively. In addition, the centrarchids were expected to show superior braking and maneuvering due to the location of the paired fins (Harris, 1953). Strike targets and angles should not depend on predator morphology and would not be expected to vary among species.

The four predators struck prey in the same region of the body, near the center of mass. Thus all struck at the same target area on the prey. Strikes were usually made from the side of the prey, anticipating the probable escape path of the target. The non-esocid predators chased 52% of fleeing prey while tiger musky chased no prey that tried to escape. The centrarchids used their paired fins for better maneuverability. Therefore, behavior was consistent with expectations derived from knowledge of relationships between body form and locomotion.

3.3 Evaluating Behavior

Although qualitative expectations of predator behavior were supported by the comparative experiments, the quantitative predictions were rejected. All the predators were expected to swim at maximum speeds in chasing prey, but only *Esox* performed maximally. Strikes by the non-esocid predators were relatively slow, with speeds typically less than 50 cm/sec. Chases were also relatively slow and short, typically less than 25 cm and lasting about 2 sec. Thus the highest speeds observed were about one seventh of expected maximum speeds.

The low speeds used by the non-esocid predators have far reaching consequences for theory. The mechanical properties of a predator's body and fin form only limit performance and influence prey-capture success when a predator has to swim at maximum speed or acceleration rate. However, since predator strike speeds were lower than the maximum possible, then the expectation that body form would limit predatory behavior is wrong. Instead, the observed behavior is one of many that are possible within the limits imposed by the body and fin form. Under these circumstances, knowledge of the locomotor mechanics and performance limits alone cannot be used to predict strike behavior.

However, prey capture effectiveness also varied between the esocid and non-esocids. Musky captured about 50% of responding prey while non-esocids captured about 15%. Chases by the latter brought the total captures up to 34% of responding prey. Overall, musky captured 65-75% of all prey attacked compared to 40-50% for the other predators. Thus the escocid body and fin form is more effective than others in capturing prey in a strike, and it might be expected that other predators should converge on this form.

The question of variability in the body and fin forms of piscivores can be approached by expanding the area of inquiry to consider locomotion needs in other phases of the predation cycle and non-predatory behavior. The form characteristic of esocids may be effective in catching prey from ambush, but it is not efficient for searching. In contrast, the form facilitating search is less effective for catching prey (Webb, 1984a). This leads to the recognition in fish of the same continuum from sit-and-wait predators to wide-searching predators that has been described for tetrapods (Chapter 6). Then, while wide-searching species have low strike-success rates they have higher encounter rates, and vice versa. The various predator behavior patterns correlate with forms that have different mechanical properties favoring continuous swimming for searching or acceleration for ambushing prey (Webb, 1984a).

3.4 The Need for Complementary Disciplines

The discussion so far has considered predation only in terms of the biomechanics of locomotion. In practice, single factor approaches to any functional morphology problem are insufficient (Chapter 2). This is apparent in the example given above, because a new question is identified that could not be predicted a priori and which cannot be addressed using locomotor mechanics principles. This question concerns the differences in strike success of the four piscivores tested; why is the strike success of *Esox* higher than that of the non-esocid predators?

It would be attractive to attribute the differences in catch success among the predators to the use of apparent sub-optimal tactics (low strike speeds) of the non-esocids. In practice, analysis using principles from ethology (Dill, 1974) showed that prey have higher response thresholds to attacks by esocids compared to the non-esocids. Where vision is important in detecting predators, prey respond to the looming image of a predator approaching on an interception course. The looming effect is calculated as the rate of change of the angle of the silhouette of the predator as viewed by the prey. The value for the rate of change of this angle is directly related to the predator's speed and size, and inversely related to the distance between the predator and prey. Therefore, prey with lower response thresholds respond to more distant or slower predators of the same size. Since the prey had lower response thresholds to attacks by non-esocids compared to attacks by *Esox*, more prey escaped from non-esocids because escape attempts were more frequent and started earlier. Although the reason for the difference is not known for certain, the variation in responses most probably relates to different sensitivities of fish to various shapes. In this case, prey responses were probably affected by the cross-sectional silhouette of an attacking predator. *Esox* has a rounded cross-sectional shape while the non-esocids have oval or lenticular cross-sections. Fish have been shown to be particularly sensitive to the acute-angled components of the latter shapes in operant conditioning trials (see Webb, 1982).

The low response threshold of prey to the non-esocids also leads to the initially unexpected recognition that the duration of an interaction can be minimized for these predators by strikes at low speeds. When prey respond to the predator's silhouette, the time taken for the predator's mouth to reach the prey from a given distance increases monotonically with strike speed to a plateau of about 120 msec at predator speeds above about 50 cm/sec. After the onset of the prey's startle response, a predator has up to about 100 msec to reach prey before the prey accelerates away from the strike path of the predator's mouth (Fig. 2). Therefore, when attacking prey with low response thresholds, predators that strike at speeds below about 50 cm/sec are most likely to close with the prey before the prey can escape. Strike speeds of the non-esocid predators with low response thresholds usually attacked prey at the appropriate low speeds.

Consideration of prey response thresholds requires revision of the theory for optimum strike tactics. Arguments based on biomechanical principles and game theory for interceptors were used to predict that predators should use acceleration lunges to minimize the duration of an interaction. It is now seen that this is only appropriate for predators attacking prey with high response thresholds. Otherwise, predators should strike at low speeds to achieve the same result of reducing the time available for prey to make an escape.

4. Strengths and Weaknesses of Biomechanical Locomotor Studies of Predation

The example illustrating the use of locomotor principles for evaluating fish predator behavior shows the strengths of studying predator-prey interactions from this perspective. The major advantage is that clear expectations for optimal locomotor behavior can be derived from measured or measurable mechanics and limits of locomotor capabilities. These expectations can be experimentally tested. In addition, trials can be repeated to obtain large sample sizes to improve accuracy or to concentrate on particular parts of predation events, such as the strike, jaw movements at capture, chase decisions, etc. The ease with which trials can be repeated also provides opportunities to examine such factors as sources of variation among predators, effects of satiation and predator or prey exhaustion.

The primary disadvantage of typical locomotor analyses of predation is that events are usually staged in an arena favorable to the viewer. The conditions for experiments are usually artificial for the animals, and hence affect behavior so that preliminary tests are necessary to ensure the design of the arena is appropriate for the particular question being researched. This can be illustrated for space, which is usually restricted in experimental studies.

In the example presented above, detailed measurements of kinematics were obtained for strikes by *Esox* using movie films and video recordings. Encounters were staged in a small space (50 cm x 50 cm x 10 cm deep) in order to be able to obtain sufficient spatial resolution for analysis of movements. However, behavior and performance were also measured during preliminary experiments using larger arenas, up to 3 x 2 m. Comparisons of observed strike behavior in the large and small arenas, and with other mainly published observations on the behavior and locomotor performance, were used to determine that the smaller arena size was adequate for the experiments. Thus initial observations and results from other experiments were used to determine an arena size appropriate for analyzing strikes, and some aspects of chase behavior.

Strikes, however, take place over short distances and times. Arenas appropriate to the study of these behaviors will restrict chases and stalks that take place over larger distances and longer times. Errors of perhaps several centimeters and tens of milliseconds would be intolerable in studying strikes, but such errors may represent small percentages of total distances travelled and times taken to complete chases and stalks. Hence lower precision is acceptable in studying longer and slower behaviors and relative accuracy equal to that required to analyse strikes can be obtained using very much larger arenas. Movie and video records are still useful in studying stalks and chases, especially the latter, but observations can also be made using log books, tape recorders and event markers to record animal behavior and movements over systems with known coordinates (Major, 1978; Messenger, 1968; O'Brien, 1979). These techniques are also appropriate for many field studies (Elliott et al., 1977).

Cover, refuges, light levels etc. are also important to the behavior of predators and prey, and they are often modified or omitted from arenas to facilitate observation. The experimental biologist recognizes these factors, and considers them variables to be explored one at a time in order to determine their importance in influencing the behavior of the participants and the outcome of interactions. Nevertheless, criticisms of artificiality can never be completely refuted, and for this reason laboratory studies must be complemented by field observations, just as the latter frequently need complementary laboratory analysis.

The example of fish predators described above also focussed upon the role of locomotion in predator-prey interactions using ideas borrowed from game theory for interceptors. No consideration was given to the costs associated with the "choices" made by the predator and prey other than to recognize that the prey usually dies by failing to respond. The time minimization/maximization assumptions of interception games are probably appropriate for the strike phase, but will be tempered over the whole predation cycle depending on such factors as whether predators are foraging to minimize time or to maximize net energy gain. Thus, the analysis of costs and benefits of various behaviors which take into account search costs, encounter rates, capture costs and payoff will influence locomotor tactics. The integration of such factors with expectations from locomotor mechanics is yet to be made. However, a mechanics approach may be of value in determining relative costs of behaviors where direct metabolic measurements are difficult or impossible (Chapter 5).

5. Further Applications

Locomotion is essential to predatory and avoidance behavior of most animals. Therefore, knowledge of the physico-chemical properties and performance limits of the structures used for movement have great potential in elucidating questions of predator-prey strategies and tactics. Research on the chemical properties of locomotor systems, especially gross metabolism, has had the greatest impact to date (Bennett, 1980; Chapter 5). Studies based on aerobic and anaerobic metabolic performance have been particularly fruitful in evaluating foraging strategies among tetrapods (Chapter 6), and analogous results are possible using principles of functional morphology. Studies on the physical properties of locomotor structures used in predation have been largely explanatory, and the value of functional morphology to suggest alternative explanations and formulate testable hypotheses has not been fully utilized. A major reason for this may be that the objectives of research on locomotor mechanics and on predation are typically different.

Mechanics research has emphasized measurement of forces and performance, especially speed. Techniques such as treadmills and flumes have channeled research into questions on constant-speed rectilinear motion. In contrast, the results of predation interactions are more likely to depend on maneuver, when linear and angular acceleration rates become large and forces will be large in directions other than those traditionally measured. It was noted above that field and laboratory approaches to predation should interact. In this case, field observations show a major vacuum requiring basic biomechanics research on animal maneuverability.

The problems of studying locomotor behavior seen during predator-prey events are formidable, especially as simple replicable experiments are desirable. Terrestrial and aquatic animals can be made to move at various speeds in circles of variable radii to study centripetal acceleration. Controlled linear accelerations are probably more difficult to achieve for aquatic, terrestrial and aerial animals. Aerial combatants probably pose the greatest problems.

I suggest that a simpler approach may be initially effective. It was pointed out above that events occuring about the prey's startle response are pivotal in an interaction, influencing predator and prey motor tactics. Startle responses are extremely easy to induce and to record using devices already standard in biomechanics laboratories. Furthermore, startle responses tend to be relatively invariant maneuvers allowing for repeatable experiments. I suggest that comparative studies of the locomotor mechanics of startle responses in many taxa are not only desirable but an eminently practical first step. Measurements of startle behavior and performance define what prey do, and as shown in the example for fish, can be used to predict, and then test, how predators should behave to anticipate the prey behavior.

Comparative studies are essential because maneuverability of predators and prey would be expected to vary greatly among taxa. This variation should in turn influence strategy and tactics, especially for the predators which initiate the interactions. For example, the structure of the body and axial skeleton make a fish flexible in lateral movements. This flexibility leads to large locomotor forces in the plane of lateral movements, this usually being the horizontal plane. In contrast, some potential prey, such as *Cyclops*, swim using their antennae and move primarily in a vertical plane. Copepods escape by swimming horizontally. Do escape trajectories influence choices of zooplankton prey? Similarly, the main bending axis of the vertebral column and the location of the appendages probably make mammal and bird piscivores more maneuverable in the vertical plane, opposite to their prey. When the predator and its prey move in different planes, strike paths and escape paths will intersect only briefly, and probably for too short a time for a predator to be able to predict interception points with a reasonable probability for success. Predator behaviors based primarily on surprise seem more likely under these conditions. But, given the shape of the body and the limb distribution of mammals, is maneuver possible by rotation about the long axis? How well can birds "bank" under water? Such factors could allow strikes in a variety of planes. Alternatively, considering the usual plane of motion of fish, have flatfish found a way to surpise prey by attacking unexpectedly in the vertical plane? Should aquatic prey respond to the presence of fish predators by aligning the body at an angle to the horizontal to move quickly away from the plane of a predator attack?

Similar questions can be imagined for terrestrial and aerial systems. Terrestrial systems may be simpler because interactions usually are spatially two dimensional. Aerial interactions take place in three-dimensional space, but the range of taxa involved is relatively small. The pervasive force of gravity and low density of air in terrestrial and aerial habitats may reduce the range of behaviors possible for predators and prey.

There are also major unresolved questions concerning the locomotor capabilities and performance of predators and prey during ontogeny. At some stage in their lives, most predators are potential prey, and many predators remain prey throughout life. However, as animals grow, the habitats they occupy and the food they eat often change. As a result, physical and sensory adaptations for predator avoidance and prey capture are likely to change throughout life. In some cases, structures that facilitate prey capture may impair escape abilities. For example, certain body forms of fish maximize acceleration rate or swimming speed or maneuverability. However, high performance in any one of these locomotor areas is associated with reduced performance in the others; i.e. optimum forms are mutually exclusive for each of the three activity patterns (Webb, 1984a). Thus specialization of the body and fins for a particular locomotor activity restricts behavior based on other locomotor patterns.

The problem of changing needs for prey capture versus predator avoidance during ontogeny is likely to be greatest for fish with body and fin forms associated with high speed swimming and cruising, such as many sharks, tunas, fossil ichthyosaurs, and cetaceans. Such animals are collectively called thunniform animals. Their specializations for high speed swimming reduce acceleration rates central to escape behavior so that such animals would be expected to be particularly vulnerable to predators, especially during early stages. Specializations for particular habitats at different life history stages occur in some animals, with successive stages linked by metamorphosis. However, Wassersug and Sperry (1977) have shown that vulnerability of anurans increases during metamorphosis. This is because there is a period when locomotor organs developing for use on land impede swimming at the same time that the residual tail interferes with terrestrial movements.

Thunniform animals do not have complex life cycles so that their young, with adult-type body and fin forms that reduce escape capabilities, would face particularly heavy predation. It may be no accident that thunniform animals are large as adults. Thunniform mammals, sharks and ichthyosaurs are (or were) usually viviparous, often with some post-partum protection of the young. This would facilitate survival of young through the period where they would be particularly vulnerable due to their locomotor form. Tuna spawn in very productive waters where early growth is very fast, thereby reducing the duration of the life history stages when they would be most vulnerable to predators. Therefore, consideration of ontogenetic changes in locomotor structure and function, and their role in predator-prey interactions, may provide novel ways of considering life histories.

6. Conclusions

Study of the locomotion of predators and prey has great potential for understanding the basis and limitations of their behavior. In determining performance capabilities of animals, biomechanical studies can determine possible behavioral options, and often predict optimum behaviors. These can be tested experimentally, but the need for interpretation and further experiments based on other disciplines are nonetheless often necessary. Many techniques have been developed, and are being improved continuously, to study the mechanics and performance of the many structures used in locomotion. They could be more widely used for predator-prey performance in air, water and on land to examine evolutionary patterns and changes during ontogeny. Biomechanical methods are being modified and new methods are being developed for use in the field. This is especially critical to the future of locomotor biomechanics because the greatest current need is for basic information on locomotor activities actually seen during predator-prey interactions.

Acknowledgments

Research described in this paper was supported by National Science Foundation grants BMS75-18423, PCM77-14664, PCM80-06469 and PCM84-01650. I thank the editors and the reviewers for their useful and constructive comments.

References

Bennett, A. F. 1980. The metabolic foundations of vertebrate behavior. *BioScience* 30: 452-456.

Dill, L. M. 1974. The escape response of the zebra danio (*Brachydanio rerio*). I. The stimulus for escape. *Anim. Behav.* 22: 711-722.

Eaton, R. C., R. A. Bombardieri, and D. L. Meyer. 1977. The Mauthner-initiated startle response in teleost fish. *J. Exp. Biol.* 66: 65-81.

Elliott, J. P., I. McTaggart-Cowan, and C. S. Holling. 1977. Prey capture by the African lion. *Can. J. Zool.* 55: 1811-1828.

Harris, J. E. 1953. Fin patterns and mode of life in fishes. In *Essays in marine biology*, ed. S. M. Marshall and P. Orr, pp. 17-28. Edinburgh: Oliver and Boyd.

Heinrich, B. 1983. Do bumblebees forage optimally, and does it matter? *Am. Zool.* 23: 273-281.

Howland, H. C. 1974. Optimal strategies for predator avoidance. The relative importance of speed and manoeuvrability. *J. Theor. Biol.* 47: 333-350.

Isaacs, R. 1975. *Differential games.* Huntington, N.Y.: Krieger Publ. Co.

Major, P. F. 1978. Predator-prey interactions in two schooling fishes, *Caranx ignobilis* and *Stolephorus purpureus. Anim. Behav.* 26: 760-777.

Messenger, J. B. 1968. The visual attack of the cuttlefish, *Sepia officinalis. Anim. Behav.* 16: 342-357.

O'Brien, W. J. 1979. The predator-prey interaction of planktivorous fish and zooplankton. *Am. Sci.* 67: 572-581.

Vinyard, G. L. 1982. Variable kinematics of Sacramento perch (*Archoplites interruptus*) capturing evasive and non-evasive prey. *Can. J. Fish. Aquat. Sci.* 39: 208-211.

Wainwright, S. A., W. D. Biggs, J. D. Currey, and J. M. Gosline. 1976. *Mechanical design in organisms.* New York: John Wiley & Sons.

Wassersug, R. J., and Sperry, D. G. 1977. The relationship of locomotion to differential predation on *Pseudacris triseriata. Ecology* 58: 830-839.

Webb, P. W. 1976. The effect of size on the fast-start performance of rainbow trout *Salmo gairdneri,* and a consideration of piscivorous predator-prey interactions. *J. Exp. Biol.* 65: 157-177.

Webb, P. W. 1978. Fast-start performance and body form in seven species of teleost fish. *J. Exp. Biol.* 74: 211-226

Webb, P. W. 1980. Fast-start locomotion and the strike tactics of *Esox.* In *Advisory workshop on animal swimming,* ed. C. A. Hui, pp. 272-299. San Diego: Naval Ocean Systems Center.

Webb, P. W. 1982. Avoidance responses of fathead minnow to strikes by four teleost predators. *J. Comp. Physiol.* 147A: 371-378.

Webb, P. W. 1984a. Body form, locomotion and foraging in aquatic vertebrates. *Am. Zool.* 24: 107-120.

Webb, P. W. 1984b. Body and fin form and strike tactics of four teleost predators attacking fathead minnow (*Pimephales promelas*) prey. *Can. J. Fish. Aquat. Sci.* 41: 157-165

Webb, P. W., and J. M. Skadsen. 1980. Strike tactics of *Esox. Can. J. Zool.* 58: 1462-1469.

Weihs, D. 1972. A hydrodynamic analysis of fish turning manoeuvres. *Proc. Roy. Soc. Lond. B* 182: 59-72.

Weihs, D. 1973. The mechanism of rapid starting of slender fish. *Biorheology* 10: 343-350.

Weihs, D., and P. W. Webb. 1984. Optimal avoidance and evasion tactics in predator-prey interactions. *J. Theor. Biol.* 106: 189-206.

4 Neural Mechanisms of Prey Recognition: An Example in Amphibians

Gerhard Roth

1. Introduction: How Do Neuroethologists Study Organisms?

Reductionism has been extremely successful in life sciences and will become even more successful with modern biotechnology. However, there has always been a strong feeling that reduction of features and processes of organisms to the lowest possible (i.e., macromolecular and biochemical) level eliminates what is to be explained, namely the organism itself as a complex network of processes (cf. Roth, 1981). Representation of complexity can only be complex in itself (although it may appear as *implicit* complexity). Not accidentally, in several papers of this volume, there is a strong call for a holistic or organismic approach in dealing with complex phenomena such as predator-prey relationships involving almost all fields of biology.

I strongly favor this view both in my theoretical and experimental work. Together with David B. Wake from the Museum of Vertebrate Zoology, University of California at Berkeley, and a larger number of students and collaborators I have been carrying out a long-term research project on plethodontid salamanders that includes genetic, zoogeographic, ecological, morphological, ethological and neurobiological studies. What David Wake and I -- and, of course, many (though not all) biologists -- want to know is how an organism works *as a whole* and how and why it evolved the way it did. My experience from this work, however, is that such an organismal approach becomes extremely time-consuming and tiresome, even if one only tries to relate morphological, behavioral, neurophysiological and neuroanatomical data to feeding in but a few salamander species (cf. Roth and Wake, 1985a,b).

In this context, neurophysiology and neuroanatomy play a special role. These disciplines often seem rather isolated from the rest of organismal biology, partly due to the high technical standard they require, but also because they are basically reductionist disciplines: when neuroethology inquires into the neural basis of behavior this necessarily implies the study of the assumed basic components of behavior, nerve cells, and their morphological, physical and biochemical properties and components. As good reductionists, neuroethologists quickly lose what they want to explain, i.e., animal behavior. Another reason is that in the past most neurobiologists have strongly underestimated the complexity of brains and neural networks, even those of invertebrates and "primitive" vertebrates. Up to now, processes of perception and recognition or of sensorimotor integration are thoroughly understood at the neuronal level in no vertebrate brain. Moreover, the strategy of present-day neuroethology is to go deeper and deeper into membrane physiology, neurochemistry and neuropharmacology, i.e., further away from the subject to be explained, although many neuroethologists deeply regret this.

Nevertheless, in this chapter I attempt to present what is known about the neural mechanisms of prey recognition in amphibians and link it to feeding behavior and partly to feeding ecology. Despite the overwhelming amount of data available there is no agreement even about the basic mechanisms of neural recognition processes, and in large parts of this chapter I will deal with the difficulties concerning this problem and propose a possible solution.

Predation is an essential type of behavior both for those who want to eat and those who want to avoid being eaten (cf. Chapters 6, 8, and 9). As is stressed by Endler in Chapter 8, the first and most important step in predation and escape from predation is prey and predator recognition. The efficiency of these processes is of immediate relevance for survival, and thus the adaptive evolution of the sense organs and neural networks involved is of great importance for the understanding of organismal evolution. However, sense organs, brains and neural networks are only parts of the organism and evolve only within its global context, its opportunities and constraints. As careful studies in salamander sensory physiology and neuroanatomy involved in feeding (Rettig and Roth, 1982; Roth and Wake, 1985a,b; Wake, 1982) show, these structures and their functions are highly influenced by changes in seemingly distant parts of the organisms such as reproductive biology and respiratory morphology which, in turn, are results of a fundamental "ontogenetic repatterning" (Roth and Wake, 1985b). This is, for example, the case in the formation of a high-speed projectile tongue, frontal eyes, strong bilateral projection of the eyes to the brain and increase in brain volume in tropical plethodontid salamanders (Roth et al., 1983).

The networks and processes involved in prey recognition I will discuss here are far from being fully understood in such an organismal context, although -- as will be seen -- there is an interesting starting point for a linkage to behavior and thus to the whole organism. I will consider a number of important related questions bearing on the amphibian visual system which, however, are of general interest for predator-prey relationships in other vertebrates: (1) Are there specific receptors or "feature detectors" for various prey, or are different prey types recognized by a "multi-purpose" network ("ensemble coding")? (2) Where in the nervous system does prey recognition occur? (3) How is the information on prey characteristics used to elicit and guide the feeding behavior?

I have tried to avoid detailed experimental and technical information and discussion. The interested reader is referred to the review articles of Grusser and Grusser-Cornehls (1976), Grusser-Cornehls (1984) and Ewert (1984).

2. Feature Detectors vs. Ensemble Coding

The feeding behavior of amphibians has been a favorite subject of neurethological research dealing with the investigation of the neural mechanisms underlying behavior. The reasons for this were that both the feeding behavior and the brain of amphibians seemed to be simple enough to establish a good correlation between the observed reactions and the neural activities in certain brain centers. Special emphasis was placed on the question of the neural mechanisms of object (e.g., prey) recognition. The understanding of the process of object recognition in vertebrate brains is of general interest; and the hope was that, due to the apparent simplicity of the brain, this understanding could be achieved more easily in these animals rather than in higher vertebrates with much more complex behaviors and brains.

If we study the feeding behavior of different amphibian groups, we see that the common assumption that amphibians indiscriminately snap at everything that moves and is not too large or too small is by no means true. There are clear differences with regard to prey preferences among amphibians, and often they seem to discriminate rather precisely between prey objects of similar shape and size. We have to assume that these behavioral differences have to do with differences in the neural sensorimotor system that guides feeding behavior.

However, from a theoretical point of view it is not clear what we have to look for when we try to investigate the mechanisms of neural guidance of feeding behavior. Can we expect that, for example, prey recognition is carried out by single "feature detector cells" such as "bug", "fly", or "worm detector", as was assumed by some neuroethologists; or is prey recognition the function of neural networks comprising the simultaneous activity of perhaps many different types of neurons?

One common argument against the "feature detector" concept is the following. A "true" prey detector should respond more or less exclusively to just one type of prey and disregard others in order to make a reliable prey distinction possible. Since all known amphibians feed on more than just one type of prey and most of them have a rather large prey spectrum, we would then have to postulate the existence of a number of different "prey detectors" equal to the number of prey types eaten by the species under consideration. Furthermore, one type of prey, e.g., a worm, may appear under very different aspects regarding shape, orientation, velocity and movement pattern, depending on how it actually moves and how the predator looks at it. Therefore, for any given type of "prey detectors" we would have to assume the existence of many different detector subtypes for any of those aspects. Thus, the number of "prey detectors" to be postulated would increase beyond any reasonable level.

Secondly, as will be shown later, in the optic tectum and diencephalon of amphibians, cells with selective response properties to just one type of prey or artificial prey stimulus are very rarely found. Most visual neurons have "broad-band" characteristics in that they respond to a certain quantitative range of an object feature. We, therefore, have to assume that prey recognition is not the function of narrowly tuned "detectors" but of an ensemble or network of cells with partly overlapping response properties that function as analyzers of certain features of prey objects such as edges, contrast, velocity, movement pattern, etc.

If we try to investigate those "recognition networks" we encounter a variety of experimental difficulties, of which the following is the most important: In contrast to the situation that would occur if true "prey detector cells" existed, we cannot expect to find immediate correlations between behavior and neural responses if prey recognition is the result of simultaneous activity of a certain number of different cells;

the similarities between neural activity and behavioral prey preferences could be "hidden" within the activities of the cell components. The most promising experimental strategy, in this case, is to conduct comparative electrophysiological studies on amphibians that differ characteristically with regard to their feeding style and their prey preferences. Regardless of how directly or remotely the activities of visual neurons may be related to the observed behavior, we should be able to find characteristic differences among the response properties of either single cells or neural networks among animals with different behavior.

On the basis of the neuroethological work of some colleagues, especially Grusser and Grusser-Cornehls (cf. Grusser and Grusser-Cornehls, 1976; Grusser-Cornehls, 1984) and Ewert and co-workers (cf. Ewert, 1984), and of my own work, which I have done partly in collaboration with W. Himstedt, I will deal here with two types of anurans, the frogs *Rana temporaria* and *Rana esculenta* and the toad *Bufo bufo*, and two urodeles, the fire salamander *Salamandra salamandra*, and the tongue-projecting plethodontid *Hydromantes italicus* (Table 1).

If we compare the feeding behavior of these four types of animals, we find great similarities between the toad and the fire salamander on the one hand, and the frog and the tongue projecting salamander *Hydromantes* on the other. The toad and the fire salamander represent the "hunter" strategy, they actively search for prey; they prefer elongate prey items, often of considerable length, such as earthworms, that usually move slowly and more or less continuously. Their ability to hit a prey on the first strike is rather poor, and they do not make much use of their tongue. I will call this the "worm preference" type.

In contrast, the frogs and the tongue projecting salamander represent the "ambush" strategy. They mostly sit and wait until a prey comes close, and then by means of extensive use of the tongue, in the

TABLE 1: Behavioral Prey Preferences

	Prey Shape	Velocity	Movement Pattern	Feeding Strategy
BUFO	Wormlike	Slow/medium	Continuous	Hunting
RANA	Small compact	Medium/fast	Jerky	Ambush
SALAMANDRA	Wormlike	Slow/medium	Continuous	Hunting
HYDROMANTES	Small compact	Medium/fast	Jerky	Ambush

frog often combined with a leap, they catch that prey. Both animals prefer small, compact prey with medium or high speed and jerky movement. They have very well developed depth perception that enables them to catch their prey mostly on the first strike. I will call this the "fly preference" type. In the following we will compare the activities of visual neurons in these four amphibians.

3. The Retino-Tectal Recognition System

Within the amphibian visual system (Fig. 1) the first step of processing information occurs in the retina. The early investigators concluded from their experiments that the retina, at the level of the retina ganglion cells (r.g.c), already performs the decisive steps of prey recognition. The retina ganglion cells, in their opinion, were specific

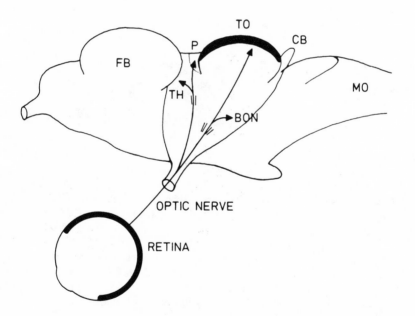

Fig. 1. Schematic drawing of the amphibian visual system. From the eye the fibers of the retina ganglion cells, constituting the optic nerve and tracts, project to thalamic (TH) and pretectal (P) areas in the diencephalon, the optic tectum (TO) and to the basal optic nucleus/neuropil (BON) in the midbrain. FB - forebrain; CB = cerebellum; MO = medulla oblongata.

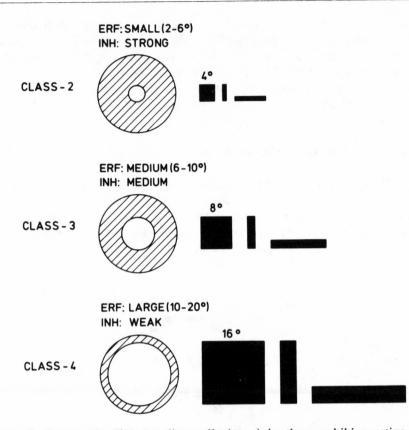

Fig. 2. Classes of retina ganglion cells (r.g.c.) in the amphibian retina. All amphibian studies so far have identified three basic types of r.g.c.: a first type (in toads and frogs called "class-2" r.g.c.) with a small excitatory receptive field (ERF) of 2-6° and a strong surrounding inhibitory zone (IRF, hatched area); a second type ("class-3" r.g.c.) with a medium-sized ERF (6-10°) and medium surrounding inhibition; and a third type ("class-3" r.g.c.) with a large ERF (10-20°) and weak or no surrounding inhibition. These types mostly respond to stimulus edges perpendicular to the direction of movement and to some degree disregard the length of an elongate stimulus. Combined with the respective optimal stimulus size of the r.g.c., this results in the stimulus preferences indicated on the right side by the order of black stimuli.

"feature detector" cells, dependent on their receptive field properties. Since one type of retina ganglion cell in the frog shows a preference for small, round, moving objects, this unit was regarded by the authors as a "bug detector". In contrast, types with large receptive fields were considered to be "enemy" or "dimming" detectors (Lettvin et al., 1959).

Later extensive studies on amphibian r.g.c. showed, however, that the differences in response properties of r.g.c. are by no means sufficient to explain the observed prey preferences in amphibian behavior. First of all, the response properties of r.g.c. proved to be highly uniform among all amphibians studied so far, regardless of their behavioral differences; secondly, the response properties of the different types of r.g.c form no distinct classes, but overlap to a great extent. In all four amphibian groups mentioned above we find three types of retina ganglion cells (Fig. 2): the first shows a small excitatory center surrounded by a strong inhibitory zone. This type responds best to small compact objects of a diameter of 2-6°. It is identical with class-1 and -2 cells in the toad and frog. The second type has a medium-sized excitatory center with a weaker inhibitory surround. It responds best to objects of medium size with a diameter of 6-10°. It corresponds to class-3 cells in anurans. The third type has a large excitatory center with rather weak or no inhibitory surround and is activated best by larger objects with a size of 10-20°. It is identical with class-4 cells in anurans (Grusser and Grusser-Cornehls, 1976; Ewert and Hock, 1972; Grusser-Cornehls and Himstedt, 1973).

As we can see in Fig. 2, the three types not only overlap in their responses to stimulus size and show the same stimulus preference order, but they respond mostly to the vertical extension of stimuli and "ignore" to a larger degree their horizontal extension. Therefore, these types cannot act as reliable prey discriminators. We have to conclude from this that the retina represents a rather unspecific filter, the properties of which cannot account for the behavioral differences.

In contrast to the situation found in the retina, where a rather limited number of response types occurs, in the optic tectum we find a large variety of response types differing with respect to their preferences of stimulus size, shape, velocity, receptive field size, movement direction, etc. However, we can reduce this variety of response types if we consider only the preferences for stimulus size, shape and velocity in the range of natural prey. If we confront the visual system with three different stimuli, a square, a horizontal bar and a vertical bar (the simplest stimuli to test visual response properties), moving at three different velocities as indicated in Fig. 3, then the following major response types are recorded (Roth, 1982; Roth and Jordan, 1982) (Fig. 4):

1. A first and always rather frequent type shows the preference S>H>V, which means that the square (S) is preferred to the horizontal bar (H) and this stimulus to the vertical bar (V). This preference is invariant with regard to a change in velocity. According to the classification of Grusser and Grusser-Cornehls (1976) and Ewert and von Wietersheim (1974) in frogs and toads this type is called T 5.1.

2. A second type also prefers S, but now responds better to V than to H at all velocities (S>V>H). In anurans this type is called T 5.3.

3. A third type is fundamentally different from the previous ones in that now it prefers H to S and to V. Again, this preference is independent of velocity (H>S>V). This type is called T 5.2.

4. A fourth type always prefers S to H and V, but prefers V at low velocity, and H at high velocity (S>VXH).

5. A fifth type also shows a preference inversion together with a change of velocity, now between H and S, such that H is preferred at low velocity, but S at high velocity. V is always the least effective stimulus (HXS>V).

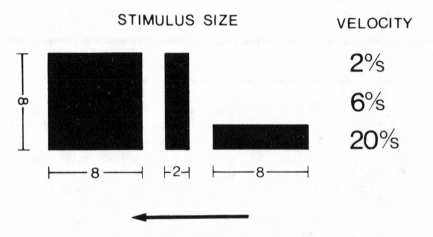

MOVEMENT DIRECTION

Fig. 3. Stimulus program for single-cell recordings in the optic tectum. The cells are activated by a square measuring 8 X 8° and a bar measuring 2 X 8° oriented perpendicular and parallel to the direction of movement. The stimuli are moved at velocities of 2, 6 and 20° per second.

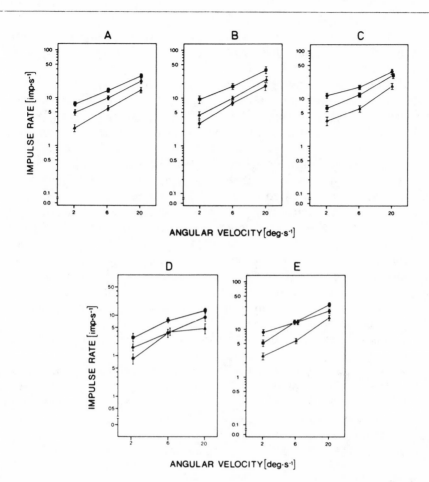

Fig. 4. Five major response types found in the optic tectum using the stimulus program shown in Fig. 3. Square symbols: square stimuli; dots: horizontal rectangles; triangles: vertical rectangles. A: type S>H>V (T 5.1); B: type S>V>H (T 5.3); C: type H>S>V (T 5.2); D: inversion type S>VXH; E: inversion type HXS>V. Figures 4A, B, C and E are taken from *Bufo* (Roth and Jordan, 1982), figure 4D is taken from *Hydromantes* (Roth, 1982). For further explanation see text.

I will disregard here all other types found under the same experimental conditions that represent together less than 20% of all tectal response types found so far.

The neurons belonging to these major response types have mostly small (6-15°) or medium (16-30°) receptive field sizes, but some also show large (31-60°) or very large (>60°) receptive field sizes. We see that at least the neurons with small or medium receptive fields are "edge (+ contrast) analyzers" in that they respond best to edges oriented either perpendicular (e.g., T 5.3) or parallel (e.g., T 5.2) to the direction of movement (the "Ewert" principle, as I will call it) or to both vertically and horizontally oriented edges (e.g., T 5.1) then constituting a "size analyzer". In some types, analysis of edges is combined with velocity preference which constitutes the preference types S>VXH and HXS>V. It is clear that with more complex stimulus situations we can find more complex combinations of different analytic tasks, e.g., a combination of edge, velocity and direction analysis. It is important, however, that even these "analytic" properties of tectal neurons are -- as can be seen in Fig. 4 -- by no means sharply tuned but broad-banded.

Let us now compare the distribution of these tectal response types in the four amphibian groups (Table 2).

We find that some types are present in all four amphibian groups: T 5.1, T 5.3, and -- at least in *Bufo*, *Salamandra* and *Hydromantes* -- the first preference inversion type (S>VXH). These types are present at different proportions. Whereas in *Bufo* and *Rana*, T 5.1 is more numerous than T 5.3, in the two salamanders the opposite is the case, especially in *Hydromantes*. We can consider these types as "basic" or "universal" response types of the amphibian tectum.

TABLE 2: Distribution of Tectal Response Types

Species	S H V T 5.1	S V H T 5.3	H S V T 5.2	S/H V	S VXH	HXS V
BUFO	26%	10%	19%	—	17%	12%
SALAMANDRA	18%	26%	15%	—	6%	9%
HYDROMANTES	15%	32%	—	—	20%	—
RANA	23%*	13%*	—	21%*	?	?

*Calculated on the basis of data from Schurg-Pfeiffer and Ewert (1981).
? = Not measured in Rana.

On the other hand, there are remarkable differences among the species: those neural response types that prefer H at all velocities or at least at low velocity, as T 5.2 or the second inversion type, are present only in *Bufo* and *Salamandra*. In *Hydromantes* both of these types are completely absent. In *Rana* a T 5.2 response characteristic is present only at stimuli of short edge length but not at 8°; here a larger number of neurons does not discriminate between S and H, as indicated in the fourth column of Fig. 4.

This shows that the presence of neural types responding best to "horizontal" edges corresponds well to the behavioral preference for wormlike prey objects and that the absence of those neural types corresponds to a behavioral preference for compact, small objects. Does this finding not revive the "feature detector" concept, especially with regard to T 5.2 neurons?

Such an interpretation would meet a variety of difficulties. First of all, even those amphibians that do not possess the "worm detector" type T 5.2, as it was often called, often snap at wormlike stimuli, though they prefer small compact prey to elongate objects when they have the choice. Furthermore, all amphibians tested so far snap even at vertically extended rectangles when they are motivated enough, though there are no "detector cells" for such stimuli. Therefore, the behavioral response to those stimuli cannot be mediated by just one type of neuron.

Secondly, as has been already mentioned above, all T 5 neurons have rather broad response characteristics and cannot, by themselves, yield a reliable distinction between different types of prey. For example, the T 5.2 neurons alone cannot reliably distinguish between compact and horizontally elongate stimuli of certain extensions, i.e. within a wider range of area dimensions for a certain "wormlike" object there is always a compact object of similar efficacy. A reliable distinction, therefore, is possible only through the simultaneous activity of different types of T 5 cells.

Thirdly, one would expect that such "command neurons" are at the top of a certain network hierarchy such that they have no primary input from retinal fibers but get their input only from other tectal neurons and represent a more or less direct input to the motor system, as has been assumed by Ewert in his model for prey recognition in the toad (cf. Ewert, 1984). However, there is no anatomical evidence for such a hierarchy. All cells belonging to the T 5 class involved in prey recognition seem to have direct retinal input (Lazar et al., 1983; Ingle, personal communication). This also contradicts the assumption that the T 5.2 characteristics ("worm-preference") are constituted by subtractive interaction between tectal and pretectal cells as is assumed by Ewert.

4. The Constitution of Tectal Response Properties

An der Heiden and Roth (1983) have developed a mathematical model for the constitution of the different response types in the optic tectum. They were able to show that only two neural principles are necessary for the formation of all observed tectal prey preference types: (i) summation of the activity of the different retinal response types, especially class-2 and class-3; and (ii) lateral recurrent inhibition among tectal neurons.

Indeed, as shown in Fig. 5, some tectal response types can be understood as the result of a simple summation process of retinal activity: in this way, the first type, T 5.1 or S>H>V, is constituted by a summation of the activity of r.g.c. class-2 and the second type, T 5.3 or S>V>H, by a summation of the activity of r.g.c. class-3.

For the constitution of T 5.2 cells that prefer horizontal bars, however, we have to assume that between tectal cells that either sum up the activity of retina class-2 cells or class-3 cells, lateral recurrent inhibition occurs, in the sense that cells, when activated, suppress the activity of their neighboring cells. This occurs either through direct inhibitory coupling or via inhibitory interneurons. Both mechanisms may occur simultaneously in the tectum.

If we vary the level of inhibitory coupling between different "first order" tectal cells we get an interesting continuum of response types: for example, if we have tectal cells summing up the activity of retina class-3 cells with no recurrent inhibition, then we get tectal type T 5.3. If there is a medium degree of recurrent inhibition, then T 5.1 arises, and with strong recurrent inhibition T 5.2 arises. This shows that we do not have to assume discrete T 5 response types in the optic tectum of amphibians. It may well be that all belong to one or two basic types, e.g., one constituted by summation of class-2 r.g.c. and the other by summation of class-3 r.g.c., which are modified into the observed response types by different degrees of lateral inhibition. Such an assumption could explain the frequent existence of intermediate response types. However, since we do not observe a real continuum of preference types, but a distribution of accumulation points, we have to postulate that there is some nonlinearity in the realization of lateral inhibition. The most important point, however, remains that it is unnecessary to assume that discrete response classes exist in the tectum.

This model also enables us to reconsider the so-called "TP-phenomenon". As shown by Ewert (1968), in the toad a destruction of the pretectal region leads to a loss of "worm/anti-worm" discrimination ability in prey-catching behavior, i.e., the ability to distinguish between elongate stimuli oriented parallel or perpendicular to the direction of movement. Ewert and co-workers observed that after such lesions in the toad's tectum T 5.2 characteristics vanished and transformed into a response property similar to the thalamic response type TH 3, which responds best to large squares and to vertically oriented bars. Additionally, the receptive field sizes enlarged from about 27° to more than 40°. The same "TP-phenomenon" was observed by Finkenstadt and Ewert (1983) in the fire salamander, here with an influence on the T 5.1 response type.

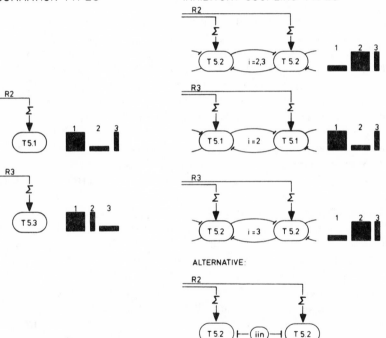

Fig. 5. Constitution of tectal response types according to the model developed by an der Heiden and Roth (1983). On the left response types constituted by mere summation of activity of retina ganglion cells (r.g.c.) are given, on the right response types constituted by summation of r.g.c. plus lateral recurrent inhibition between tectal cells are shown. (i) indicates strength of inhibitory coupling expressed in arbitrary units. Row of black stimuli indicates the respective stimulus preference. In the bottom line an alternative for inhibitory coupling through inhibitory interneurons (iin) is presented. R 2 = r.g.c. class-2; R 3 = r.g.c. class-3.

In the original version of the Ewert-von Seelen model of prey recognition in the toad, it was assumed that the T 5.2 characteristic is exclusively due to a subtractive inhibitory interaction between tectal T 5.1 characteristics and pretectal TH 3 characteristics (Ewert and von Seelen, 1974). If this were the case, then a destruction of the thalamo-pretectal region including TH 3 cells and, therefore, an elimination of their inhibitory influence, should turn T 5.2 characteristics back into T 5.1 ones. This, however, is not observed; the "disinhibited" T 5.2 cells now show preferences for very large objects ($>32°$). Furthermore, the "disinhibition" effect occurs also in those animals that do not have clear T 5.2 characteristics, as is the case in *Rana pipiens* (Ingle, 1973).

We can avoid these difficulties by assuming that the inhibitory influence on which the "TP-phenomenon" is based is a non-specific one, and that the constitution of type T 5.2 occurs through recurrent inhibition between T 5.1 cells within the tectum. There is some anatomical evidence that two types of inhibition exist within the tectum: one intrinsic through stellate cells, and one extrinsic through fibers coming from pretectal neurons (Szekely and Lazar, 1976). Szekely and Lazar discuss both types of inhibitory processes as being involved in the restriction of lateral spread of local activation over the whole tectum. However, the extrinsic inhibitory influence may consist of controlling and restricting the lateral spread of activity and thus the receptive field size of tectal cells, while the inhibitory stellate cells may produce the specific inhibitory effect among T 5 cells.

After destruction of the pretectum, disinhibition results from a general increase in receptive field size and optimal stimulus size and, therefore, loss or impairment of the response properties of T 5 neurons. Due to this increase, the "worm/antiworm" discrimination would vanish because both types of elongate stimuli are now below the optimal size level. The "disinhibition" effect would be an unspecific one and not the result of the destruction of specific network properties. However, the possibility remains that the inhibitory interneurons are themselves influenced by specific thalamo-pretectal cells. What is important is to recognize that there seems to be a general inhibitory influence of the thalamo-pretectal region on the tectum independent of the T 5.2 characteristic that controls receptive field size of tectal cells.

5. The "Recognition Module" Concept

The model I propose here for prey recognition is the following: a certain prey object activates a spatially restricted part of the optic tectum, depending on where in the visual field the object moves. This part of the tectum comprises a certain number of neurons involved in prey recognition, such as the T 5 neurons presented above. I will call this tectal space a "recognition module". It has to be emphasized that these modules are not understood as anatomically stable "columns". There is no evidence that the amphibian tectum is organized anatomically in a columnar way. The idea is that a minimum set of T 5 neurons differing in response characteristics are activated simultaneously by a prey stimulus, and that the borders of this set can shift anywhere. The properties of such a module are determined by the following:

(i) the qualitative composition of the module with respect to the different neural preference types, i.e., which of the T 5 cells are present,

(ii) the quantitative composition of the module, i.e., the numerical percentage of the different preference types within the module, and

(iii) the relative contribution of the different preference types to the overall activity of the module, depending on the activity level of the types.

Such a module, in which, among other neurons, all T 5 neurons are active simultaneously, can act as a universal "prey analyzer": any prey object will activate all components of this module, due to the fact that the response properties of the components are overlapping. The components of this module act as semi-specific analyzers with respect to single characters of the prey object as explained above. They "decompose" the objects into those characters, and their simultaneous activity represents the prey object. Accordingly, different prey objects will activate the different components differently depending on the properties of the prey and therefore elicit a distinct overall activity pattern within the module. There is no need for a large number of discrete prey detector cells in the tectum. It is interesting that similar models of "global" coding of external stimuli have been recently developed for different animals and different modalities (McIlwain, 1976; Scheich, 1983).

6. Neural Guidance of the Feeding Sequence

The process of prey recognition is the first step in the whole sequence of feeding. There has been some controversy between Grusser/Grusser-Cornehls and Ewert about the neural guidance of the feeding sequence. The question is whether prey recognition is achieved by specialized "recognition cells" (e.g., cells of the T 5 type, which are active during the whole feeding sequence as is claimed by Ewert [cf. Ewert, 1984]), or whether prey recognition is a function of *all* tectal cells involved in the guidance of feeding, as Grusser and Grusser-Cornehls contend (e.g., Grusser and Grusser-Cornehls, 1976). The latter concept is based on the findings of the authors that different classes of tectal cells are apparently involved in the guidance of different parts of the feeding sequence. For example, "small field" T 5 neurons, as discussed above, are involved in prey recognition; T 2 neurons, which are optimally activated by temporo-lateral movement of small objects, are responsible for turning toward prey; T 3 neurons, which respond best to objects moving toward the eye, are involved in approach; T 1 neurons, which are activated only binocularly, are responsible for binocular fixation; etc. Thus, the whole sequence of activity of T 5 - T 2 - T 3 - T 1 - T 3 neurons would lead to a behavioral sequence of recognition/turning - approach - binocular fixation - snapping.

Ewert has argued against this concept; i.e., that the "recognition elements" (T 5.2 neurons in the toad) have to be active during the whole feeding sequence. He, therefore, constructs the sequence of neural activities in the tectum in the following way: T 4 (broad-field neuron) + T 5.2 (= "stimulus moving anywhere in the visual field" + "object recognized as prey") --> ORIENT, PREY!; T 2.2 + T 5.2 + T 1.1 (= "stimulus in the frontal field" + "object recognized as prey located near the fixation area" + "stimulus far afield") --> APPROACH!; T 5.2 + T 1.2 (= "prey near the fixation area" + "stimulus close to the toad") --> FIXATE!; T 5.2 + T 1.3 + T 3 (= "prey inside fixation area" + "both retinas simultaneously stimulated" + "stimulus within snapping distance") --> FIXATE and SNAP! (cf. Ewert, 1984).

Apart from the fact that in this concept Ewert uses a further refined classification with respect to Grusser and Grusser-Cornehls (e.g., T 2.2, T 1.1), one sees that the T 5.2 activity indeed accompanies the activities of other specialized neurons. Ewert is surely right in stating that the act of recognition of an object as prey has to precede all other acts, but this has not been denied by Grusser and Grusser-Cornehls: their sequence of neural activity also begins with T 5 neurons. Ewert, however, disregards the experimental evidence that neurons of classes other than T 5 alone can have "recognition properties". Roth (1982) in *Hydromantes* and Roth and Jordan (1982) in *Bufo* often recorded from large-field neurons such as T 4, or direction-specific neurons such as T 2, which showed clear discrimination between horizontally and vertically oriented stimuli of the size of a prey. Therefore, it is not *per se* necessary to assume that the T 5 cells are always active during the feeding sequence, although this may well be the case. The argument of Ewert based on recordings from freely moving toads (cf. Ewert, 1984) is somewhat circular because all neurons showing a "worm preference" were classified as T 5.2 even though a preference for "worm-like" stimuli could be found in other cell types as well. There is no doubt, however, that this sort of experiment will give much more precise insight into the guidance of the feeding behavior.

In summary, we have to assume that the T 5 cells are necessary but not sufficient to guide the feeding sequence, regardless of whether they are active only at the beginning of or during the whole sequence. One of the still unsolved major problems consists of the neural basis of depth perception. All T 5 neurons show angular instead of absolute size constancy, i.e., they respond equally to a small and nearby object as well as to a large, distant object. Thus, for the identification of an object as prey, additional information about the distance of the object is needed. Several authors have shown that two-eyed amphibians primarily use binocular cues for depth perception, whereas one-eyed animals use accommodation (Collett, 1977; Jordan et al., 1980). There is almost no knowledge about how these processes are combined with the activity of T 5 cells.

An interesting new aspect comes from recent findings by Ingle in the frog *Rana pipiens*. Ingle (Ingle and Crews, 1985, and personal communication) was able to show that destruction of the uncrossed tecto-bulbar tract eliminates snapping but not turning toward the prey, whereas destruction of the crossed tecto-bulbar tract abolishes turning but not snapping. Ingle was able to show by means of the horseradish peroxidase (HRP) method that the uncrossed tract originates from tectal neurons that have contact with class-3 r.g.c. and the crossed tract originates from tectal neurons that have contact with class-2 r.g.c. Accordingly, after destruction of the tectal fiber layer containing fibers from class-2 r.g.c. the orienting/turning responses disappear, whereas this is the case for the snapping response after destruction of the deeper fiber layer belonging to class-3 r.g.c. Ingle deduces from this that turning towards prey (including prey recognition) involves cells activated mainly by class-2 r.g.c., whereas snapping is elicited by cells activated mainly by class-3 r.g.c. activity.

Now we have to consider the fact that class-2 r.g.c. are best activated by objects of small angular size (2-6°) and that class-3 r.g.c. by medium angular size (8-10°). Then we can assume that a prey of small absolute size activates class-2 r.g.c. when distant, and class-3 r.g.c. when near, i.e., within snapping distance. Interestingly, as was shown by an der Heiden and Roth in their model presented above, we can assume that two subclasses of T 5 cells are present in the tectum, one driven by class-2 r.g.c. and one driven by class-3 r.g.c., but showing the same response properties. Therefore, it is possible that neurons of the first subclass are active during fixation of distant prey and neurons of the other subclass during fixation of near prey. This would make it unnecessary to assume a direct influence of the binocular or monocular depth perception system onto the T 5 cells. This speculation remains to be proven by further experiments.

7. Sensorimotor Integration

A further important but still unsolved question is how and where in the brain the activity of tectal neurons involved in prey recognition is summarized and transformed into impulses activating the appropriate motor nuclei. As shown in the preceding section, several behavioral programs, such as for locomotion during approach, fixation, and estimation of distance of the prey, are needed for feeding and have to

Fig. 6 (facing page). A comparison between the functional organization of the tectum of anurans and urodeles. On the left side of the drawings the cell and fiber layers of the tectum are given according to the nomenclature of Szekely and Lazar (1976). Letters indicate superficial retinal fiber layers. Four basic types of tectal neurons can be distinguished: a) neurons with small pear-shaped somata and small dendritic trees making contact with different types of retinal afferents (SP1, SP2); these neurons represent axonless interneurons or efferent elements with short-range axons; b) neurons with large pear-shaped somata and wide dendritic trees arborizing in different layers of retinal afferents (LP, LP1, LP2) which represent convergent efferent elements of the tectum and send their axons mostly to the tegmentum tecti or to diencephalic (and prosencephalic?) centers; c) large ganglionic cells with wide and mostly flat dendritic trees (G1, G2) which constitute with their axons the crossed (G1) and uncrossed (G2) descending tracts connecting the tectum with the motor zones of the medulla oblongata and the spinal cord. The tectum of anurans and urodeles differs only in that during ontogeny the small pear-shaped cells migrate toward the tectal surface in anurans, whereas in urodeles they remain near their point of origin. The other cell types are found at corresponding sites of the tectum. The drawings are based on Szekely and Lazar (1976) and Lazar et al. (1983) for anurans and on unpublished results of Grunwald, Linke and Roth for urodeles.

ANURA

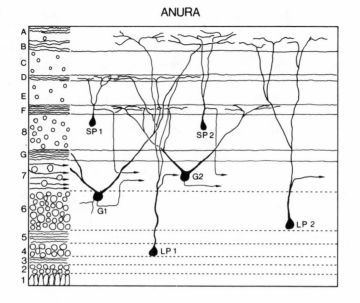

URODELA

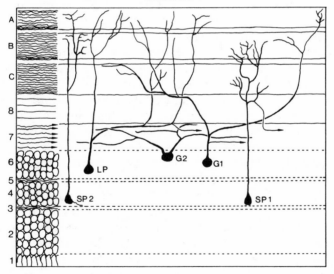

be combined with the activity of the networks involved in prey recognition. Furthermore, brain centers must exist in which information from other sensory modalities (e.g., olfactory, somatosensory, vestibular) are combined with visual information. There are four possible sites of transformation of tectal visual activity into guidance of motor action: (a) the tectum itself; (b) the peduncle and/or isthmic region in the ventral midbrain; (c) the reticular formation in the neighborhood of the motor nuclei in the medulla oblongata related to feeding; and (d) the motor nuclei responsible for the motor actions constituting the sequence of feeding, i.e., orienting, approach and snapping.

Anatomical and recent electrophysiological evidence (cf. Ewert, 1984) contradicts the assumption that the tectum is the place of *complete* transformation of visual activity involved in prey recognition into motor commands. This does not exclude the possibility that there is a high degree of interaction of different neurons involved in prey recognition or guidance of motor behavior, and that there is partial convergence, or a mixture of convergent and parallel processing, in such a way that certain neurons converge with axon collaterals on second order tectum neurons (such as pyramidal and ganglionic cells in the anuran tectum) and at the same time project themselves outside the tectum. There may be larger differences among anurans and urodeles with respect to intertectal convergence, with a perhaps higher level of such a convergence in anurans and a lower one in urodeles. The traditional view on the urodele tectum is that there is little, if any, convergence within the tectum, and that most units project themselves outside the tectum (Leghissa, 1962). Recent neuroanatomical experiments in our laboratory (Roth et al., in preparation) show that in salamanders the situation is very similar to that found in anurans: they possess the same basic types of tectal neurons both with regard to morphology and function (e.g., target site of axonal projection) (Fig. 6).

The next question is whether tectal neurons involved in prey recognition and the guidance of the feeding sequence project directly to the corresponding bulbar and spinal motor nuclei or to an intermediate "convergence center". The first possibility is contradicted by the fact that there is little anatomical evidence for such a direct projection. Experiments with selective staining of the large efferent tracts that descend from the tectum to the medulla oblongata and the cervical spinal cord show that these tracts do not massively terminate on the motor nuclei that control feeding behavior. These are the motor nuclei of cranial nerves VII, IX and X and of the first and second spinal nerves in urodeles (Roth and Wake, 1985a) and the hypoglossal nucleus

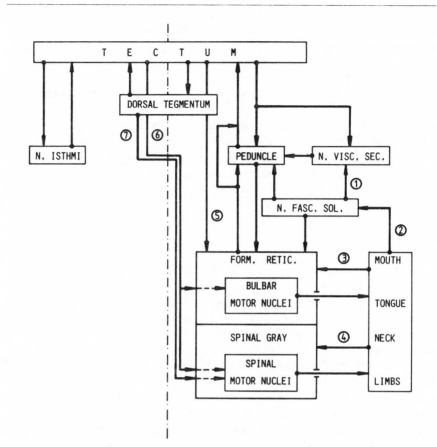

Fig. 7. Schematic drawing of descending and ascending pathways between tectum and motor zones in the brainstem and cervical spinal cord involved in the coordination of feeding. Reciprocal connections between tectum and nucleus isthmi are included. Only one half of the mirror-symmetric pathways is drawn. Abbreviations: n. isthmi = nucleus isthmi; n. visc. sec. = nucleus visceralis secundarius; n. fasc. sol. = nucleus fasciculi solitarii; form. retic. = formatio reticularis; 1 = tractus visceralis ascendens, 2 = fasciculus solitarius (visceral sensory fibers), 3 = tractus descendens nervi trigemini (somatosensory fibers), 4 = fasciculus gracilis/cuneatus (spinal somatosensory fibers), 5 = tractus tecto-bulbaris rectus (uncrossed descending tectal efferents), 6 = tractus tecto-bulbaris cruciatus (crossed tectal efferents), 7 = tractus tecto-/tegmento-spinalis (mostly crossed tectal and tegmental efferents terminating in the cervical spinal cord).

in anurans. A more theoretical argument against the assumption of direct projection of the tectal efferents to the motor nuclei related to feeding is that then either the tectum or the motor nuclei would have to be the place of convergence of *all* the other information related to feeding, such as motivation, learning, olfaction, for which there is no evidence.

The more realistic candidates for a "convergence center" are the ventral midbrain (pedunculo-isthmic region) and reticular formation surrounding the motor nuclei in the medulla oblongata and the cervical spinal cord (Fig. 7). These two regions together seem to form a two-step convergence and transformation system.

8. Summary

I have shown that in amphibians the process of prey recognition cannot be understood as the result of activity of single "prey detector" neurons: in contrast to the ability of fine discrimination between different prey types in the behaving animal, tectal as well as diencephalic visual neurons have rather broad-banded response properties. Fine prey discrimination can occur only through superposition of the activity of a certain number of types of tectal (and maybe diencephalic) neurons that respond differently to different properties of the prey stimulus such as contrast, size, orientation, and velocity, although these responses are again not sharply tuned.

I assume that during stimulation by a prey these different response types form a functional unit which I call a "recognition module". Due to its composite and analytic nature it can act as a "universal prey detector".

We can explain the formation of the different tectal response types involved in prey recognition by a combination of two processes: summation of activity of different retinal response types and lateral recurrent inhibition among tectal cells. This enables us to understand the formation of different behavioral prey preferences: different amphibians such as *Bufo, Rana, Salamandra* and *Hydromantes* show characteristic differences in the composition of their tectal "prey recognition modules" in which the presence and the degree of lateral inhibition plays a decisive role. In amphibians of the "worm-preference" type such as *Bufo* and *Salamandra*, lateral inhibition is realized to a higher degree and leads to the formation of specific response types such as T 5.2, whereas in amphibians of the "fly-preference" type lateral inhibition among specific tectal cells is weak, as in *Rana*, or absent, as in *Hydromantes*, resulting in an incomplete formation or absence of T 5.2 characteristics. The T 5.2 characteristics have to be understood as analyzers for stimulus orientation parallel to the direction of movement, not as a "worm detector". The presence or absence of the T 5.2 type will change the overall activity pattern of the recognition module and determine the behavioral prey preferences.

It is probable that in the tectum only partial convergence of the components of the "recognition module" occurs (e.g., on ganglionic cells). Other centers of convergence seem to be the isthmic and peduncular region and the reticular formation around the motor nuclei related to feeding.

There can be no doubt that very much of what was said in this paper is still speculation. However, the recent link between neurophysiology and neuroanatomy thanks to new staining methods such as microinjection of dyes during recording has strongly enhanced progress in the analysis of cooperative neuronal networks.

On the other hand, too little is still known about feeding behavior and prey preferences of amphibians in their natural environment. The analysis of stomach contents, for example, is useful, but what an amphibian eats is to a large degree determined by food availability, and real prey preferences will become recognizable only under conditions of prey abundance. Differences in feeding strategy and feeding ecology, which so far have been strongly neglected in neuroethology, are not easily worked out under laboratory conditions. Recent behavioral and ecological studies on feeding in plethodontid salamanders showed that the major differences between North American Plethodontini with "flipping" tongue and tropical Bolitoglossini with a free, projectile tongue concern not so much *what* these animals eat, but *where* they eat it: only the Bolitoglossini with their projectile tongues can feed from any three-dimensional (even upside down) position, which is absolutely necessary for arboreal life (Wake and Roth, in preparation). Apparently, all other salamanders are able to feed only in a horizontal or slightly oblique plane. These differences in feeding ecology necessarily determine the evolution of physiological and neuronal mechanisms related to feeding, although they may not easily become evident in simple stimulus situations such as used in neuroethological work. A true link between ecology and neuroethology of feeding is still out of reach.

References

An der Heiden, U., and G. Roth. 1983. Cooperative neural processes in amphibian visual prey recognition. In *Synergetics of the brain*, ed. E. Basar, H. Flohr, H. Haken, and A. J. Mandell, pp. 299-310. Berlin: Springer-Verlag.

Collett, T. 1977. Stereopsis in toads. *Nature* 267: 349-351.

Ewert, J.-P. 1968. Der Einfluss von Zwischenhirndefekten auf die Visuomotorik im Beute- und Fluchtverhalten der Erdkrote (*Bufo bufo* L.). *Z. vergl. Physiol.* 61: 41-70.

Ewert, J.-P. 1984. Tectal mechanisms that underlie prey-catching and avoidance behaviors in toads. In *Comparative neurology of the optic tectum*, ed. H. Vanegas, pp. 247-416. New York: Plenum Press.

Ewert, J.-P., and F. Hock. 1972. Movement sensitive neurones in the toad's retina. *Exp. Brain Res.* 16: 41-59.

Ewert, J.-P., and W. von Seelen. 1974. Neurobiologie und Systemtheorie eines visuellen Erkennungsmechanismus bei Kroten. *Kybernetik* 14: 167-183.

Ewert, J.-P., and A. von Wietersheim. 1974. Musterauswertung durch tectale und thalamus/praetectale Nervennetze im visuellen System der Krote (*Bufo bufo* L.). *J. Comp. Physiol.* 92: 131-148.

Finkenstadt, T., and J.-P. Ewert. 1983. Visual pattern discrimination through interactions of neural networks: A combined electrical brain stimulation, brain lesion, and extracellular recording study in *Salamandra salamandra*. *J. Comp. Physiol.* 153: 99-110.

Grusser, O.-J., and U. Grusser-Cornehls. 1976. Neurophysiology of the anuran visual system. In *Frog neurobiology*, ed. R. Llinas and W. Precht, pp. 298-385. Berlin: Springer-Verlag.

Grusser-Cornehls, U. 1984. The neurophysiology of the amphibian optic tectum. In *Comparative neurology of the optic tectum*, ed. H. Vanegas, pp. 211-245. New York: Plenum Press.

Grusser-Cornehls, U., and W. Himstedt. 1973. Responses of retinal and tectal neurons of the salamander (*Salamandra salamandra*) to moving visual stimuli. *Brain Behav. Evol.* 7: 145-168.

Ingle, D. 1973. Disinhibition of tectal neurons by pretectal lesions. *Science* 180: 422-424.

Ingle, D., and D. Crews. 1985. Vertebrate neuroethology: definitions and paradigms. *Ann. Rev. Neurosci.* 8: 457-494.

Jordan, M., G. Luthardt, C. Meyer-Naujoks, and G. Roth. 1980. The role of eye accommodation in the depth perception of common toads. *Z. Naturforsch.* 35C: 851-852.

Lazar, G., P. Toth, G. Csank, and E. Kicliter. 1983. Morphology and location of tectal projection neurons in the frog: A study with HRP and cobalt-filling. *J. Comp. Neurol.* 215: 108-120.

Leghissa, S. 1962. L'evoluzione del tetto ottico nei bassi vertebrati. *Arch. Ital. Anat. Embriol.* 67: 343-413.

Lettvin, J. Y., H. R. Maturana, W. S. McCulloch, and W. H. Pitts. 1959. What the frog's eye tells the frog's brain. *Proc. Inst. Radio Engrs.* 47: 1940-1951.

McIlwain, J. T. 1976. Large receptive fields and spatial transformations in the visual system. In *International review of physiology, Neurophysiology II,* Vol. 10, ed. R. Porter, pp. 223-248. Baltimore: University Park Press.

Rettig, G., and G. Roth. 1982. Afferent visual projections in three species of lungless salamanders (Fam. Plethodontidae). *Neurosci. Lett.* 31: 221-224.

Roth, G. 1981. Biological systems theory and the problem of reductionism. In *Self-organizing systems*, ed. G. Roth and H. Schwegler, pp. 106-120. Frankfurt: Campus Verlag.

Roth, G. 1982. Responses of the optic tectum of the salamander *Hydromantes italicus* to moving prey stimuli. *Exp. Brain Res.* 45: 386-392.

Roth, G., W. Grunwald, R. Linke, G. Rettig, and B. Rottluff. 1983. Evolutionary patterns in the visual system of lungless salamanders (fam. Plethodontidae). *Arch. Biol. Med. Exp.* 16: 329-341.

Roth, G., and M. Jordan. 1982. Response characteristics and stratification of tectal neurons in the toad *Bufo bufo* (L.). *Exp. Brain Res.* 45: 393-398.

Roth, G., and D. B. Wake. 1985a. The structure of the brainstem and cervical spinal cord in lungless salamanders (family Plethodontidae) and its relation to feeding. *J. Comp. Neurol.*, in press.

Roth, G., and D. B. Wake. 1985b. Trends in the functional morphology and sensorimotor control of feeding behavior in salamanders: An example of the role of internal dynamics in evolution. *Acta Biotheor.*, in press.

Scheich, H. 1983. Sensorimotor interfacing. In *Advances in vertebrate neuroethology*, ed. J.-P. Ewert, R. R. Capranica and D. J. Ingle, pp. 7-14. New York: Plenum Press.

Szekely, G., and G. Lazar. 1976. Cellular and synaptic architecture of the optic tectum. In *Frog neurobiology*, ed. R. Llinas and W. Precht, pp. 407-434. Berlin: Springer-Verlag.

Wake, D. B. 1982. Functional and developmental constraints and opportunities in the evolution of feeding systems in urodeles. In *Environmental adaptation and evolution*, ed. D. Mossakowski and G. Roth, pp. 51-66. Stuttgart: Gustav Fischer Verlag.

5   Measuring Behavioral Energetics

Albert F. Bennett

1. Introduction

Energy is often regarded as the primary medium of biological exchange between an organism and its environment (e.g., see Townsend and Calow, 1981). Even if energy is not seen as the ultimate determinant of biological interactions, it is surely a factor of significance that an animal must take into account. From a theoretical perspective, an animal may be expected to maximize net energy intake and minimize energy expenditure, thereby maximizing growth and/or reproduction. During predation, for instance, a predator might attempt to capture a prey if the product of the net energetic gain available and the probability of capture exceeds the cost of the capture attempt. As prey are almost always energy-rich in comparison to the behavioral cost of obtaining them, we might anticipate that a predator should always make a capture attempt if only energetic grounds are considered. In comparison, a prey organism, standing to lose everything by being consumed, should be willing to commit any amount of energy to escape successfully.

Such theoretical predictions are not terribly helpful to our understanding of behavioral interactions. In particular, they do not recognize intrinsic (i.e., physiological or anatomical) limitations on energy expenditure in different kinds of organisms (Bennett et al., 1984; Chapters 2, 4, 6). The metabolic systems of some animals, particularly ectothermic animals, place strong constraints on rates of energy utilization and constrain potential behavioral responses (Bennett, 1980; Pough, 1983). Only by actual measurement of energy expenditure can we know the real cost that animals can and do commit to a behavioral interaction such as predation. This essay examines methodologies available for making such experimental observations. These are discussed with reference to their practicality under both field and laboratory conditions.

The discussion of this topic is divided into the measurement of the energetics of sustainable behaviors, those that can be maintained for long periods of time, and non-sustainable behaviors, in which the animals rapidly tire and sometimes exhaust. This distinction is made because both the methodologies involved and the ecological questions of interest are different for the two intensities of activity. The level of exertion or work that distinguishes sustainable from non-sustainable behavior varies greatly among animals as a function of their capacity for oxygen consumption; sustainable behaviors must be supported aerobically. For instance, sustainable running speeds of mammals are nearly ten times greater than those of lizards (Bennett and Ruben, 1979; Garland, 1982), a reflection of the substantially higher levels of maximal oxygen consumption of the former. When aerobic capacities are exceeded, supplementary anaerobic metabolism is activated. Anaerobic metabolism, for reasons not yet understood, results in rapid tiring and exhaustion. Thus, these behavioral distinctions are largely determined by the type of metabolic energy utilization. The methods for measuring aerobic and anaerobic metabolism are very different and will be detailed in the next two sections.

The questions of energetic interest also are different for these two metabolic modes. For either aerobic or anaerobic metabolism, it is possible to stipulate the cost of a behavioral act, either in joules or high energy phosphate equivalents. For meaningful biological interpretation, this cost must be placed within a broader energetic context for the organism. For aerobic metabolism, it seems to me that there are two important contexts: the total energy budget and the maximal oxygen consumption. In the former case, it is possible to stipulate how expensive a behavior is to the animal as a percentage of its daily energy expenditure. Does it represent a major drain on energy resources or is it a trivial expense? If it is expensive, we might expect physiological, morphological, and/or behavioral adjustments to minimize those costs. If it is cheap, such adaptations might not be expected, at least on energetic grounds alone. Comparisons of actual with maximal oxygen consumption permit us to determine the percentage of an animal's aerobic scope that the behavior entails. If the behavior requires oxygen consumption close to the upper limits of delivery, the animal runs the risk of activating anaerobic metabolism with its ensuing exhaustion. If rates of oxygen consumption are nearly maximal, the behavior may be an important factor in maintaining maximal oxygen consumption at its present level.

Because of its brevity, anaerobically-supported behavior does not make a significant impact on an animal's total energy expenditure. Even though it may involve rates of substrate utilization and high energy phosphate formation far higher than those of aerobic metabolism (see Bennett, 1978; Gatten, 1985), it is undertaken for such brief periods that it has little significance in terms of an energy budget. The only exceptions are exposure to long-term hypoxia or anoxia, and these are only arguably "behaviors." The questions of interest in regard to anaerobic metabolism are whether it is activated at all and whether it approaches the total possible anaerobic energy delivery (the anaerobic capacity, as defined by Bennett and Licht [1972]) with its consequent exhaustion. We might expect behaviors to be structured such that anaerobiosis might be avoided altogether. If intense activity is required and anaerobic metabolism must be activated, we might expect the behavior to be limited in duration and intensity such that potential exhaustion is carefully avoided.

Although the focus of this symposium is on predator-prey interactions, the techniques discussed here are applicable to a variety of other behavioral activities, such as territorial defense, courtship, or foraging. Hence, the scope of this article is more all-inclusive regarding the measurement of behavioral costs in general.

2. Measuring the Energetics of Sustainable Behaviors

Behavioral energetics may be determined either under natural field or controlled laboratory conditions. Both have their advantages and difficulties. From an ecological point of view, it is always more desirable to undertake measurements under field conditions, where an animal is exposed to its natural environment. However, this situation may place impossible demands on analytical techniques. The artificiality of the laboratory environment must be balanced against opportunities to undertake carefully controlled measurements of energy exchange.

Several methods have been used to determine the costs of sustainable behaviors under field conditions. These involve either direct or indirect measures of gas exchange or measurement of variables correlated with gas exchange. It is generally impractical to collect expired gases from an animal active under field conditions without grossly disrupting its behavior. This has been done successfully on sea turtles by Jackson and Prange (1979), who were able to measure the cost of egg laying and of routine locomotion in situ on tropical beaches. This was an impressive achievement, but probably says more about the single-mindedness of sea turtles than about the general applicability of the method.

In recent years, gas exchange under field conditions has been successfully measured indirectly with isotopically-labelled water. Injected isotopes of hydrogen and oxygen are removed over time from an animal's body fluids and metabolic rate can be calculated from their differential elimination (see Nagy [1975] for a discussion of the technique). The use of doubly-labelled water has provided considerable information on daily or weekly rates of energy utilization by free-ranging animals. However, the technique generally lacks the resolution required for measurement of the cost of discrete bouts of behavior. There is an insufficient decrease in isotopic concentrations over the short time courses of most behaviors of interest, particularly in ectothermic animals with low metabolic rates. In addition, because the technique measures carbon dioxide production rather than oxygen consumption, measurements over short time intervals can be greatly influenced by respiratory alkalosis or metabolic acidosis. Measured values would not in these instances indicate true metabolic rate. There would, of course, be great problems in obtaining samples of body fluids immediately before and after a desired behavior. Finally, the technique is inappropriate for most aquatic organisms because of the rapid rates of isotopic loss. Its use, therefore, in measuring behavioral energetics is necessarily limited to situations involving prolonged behaviors and high metabolic rates. It has been used to measure the cost of flight in birds (e.g., Flint and Nagy, 1984; Lefevre, 1964) and might be useful to estimate foraging costs of a widely ranging animal over an entire day or night.

Various attempts have been made to estimate metabolic costs by telemetering physiological variables correlated with oxygen transport, e.g., heart rate or breathing rate. Such variables have the advantage of being easily detected in free-ranging animals, and simultaneous visual observations of behavior would permit an estimation of energetic cost. The problem with the method is the lack of tightness of the association of the measured variable and metabolic rate. A close positive association between a variable and oxygen consumption is not sufficient if a large amount of residual variability remains unexplained. The relationship may vary considerably among individual animals and even within a single individual at different times (Hargrove and Gessaman, 1973; Johnson and Gessaman, 1973). On the whole, this approach is not promising. It requires extensive validation experiments prior to field use without any guarantee that the method can in fact be employed.

None of the previously described techniques holds out much hope for permitting determination of behavioral energetics under field conditions. At present, the outlook for successful and unambiguous studies is bleak unless new methods are developed.

The costs of sustained activity under laboratory conditions are measured by oxygen consumption during either locomotion on a treadmill (or wind tunnel or flow tank) or a staged behavioral event. Both approaches have been very successful in accumulating data on behavioral energetics.

Oxygen consumption increases as locomotor speed increases. Gas exchange is measured by the change in oxygen partial pressure in the respiratory medium as it passes by the animal. The exact form of the relationship between metabolic rate and speed depends on body size, medium, and locomotor mode (see Schmidt-Nielsen, 1984). It is possible to predict from general allometric equations the cost of moving at any speed knowing only the type of animal in question and its mass (see, for example, Taylor et al., 1982). If greater accuracy is desired, experimental measurements can be made on the species or even the individuals of particular interest. Such measurements, made under steady state conditions, should accurately reflect the costs of such sustained behaviors as fish migration or bird flight. Estimated costs may also be used in theoretical models to evaluate the energetic significance of a particular locomotor behavior (e.g., see Garland [1983] regarding the cost of foraging in mammals). The only difficulty in applying this approach is that it is limited to forward locomotion at constant speed, and many behaviors do not take this form.

A second approach, which is less dependent on steady state locomotor activity, is the direct measurement of oxygen consumption during a behavioral event. This is usually done with the animal or animals in a chamber that is big enough to permit unrestrained behavior but small enough to permit a measurable decrement in oxygen partial pressure. If the chamber is completely sealed, samples of the medium can be removed immediately before and after the behavior and analyzed for total oxygen consumption during the event. In this case, care should be taken to avoid hypoxia, and instantaneous rates of oxygen consumption are not obtained. If the medium is monitored and recirculated through the sealed chamber, hypoxia remains a potential problem but instantaneous rates can be determined (Bartholomew et al., 1981). A flow-through system avoids potential hypoxia, but requires a sufficient decrement in oxygen partial pressure to permit accurate determination of oxygen consumption. This situation often requires such low flow rates that the response of the system is too slow to permit discrete analysis of metabolic costs. A closed system is usually recommended.

The chamber approach has obvious limitations that exclude behaviors of animals of large body size (e.g., cost of combat by male elk) or behaviors that require considerable space (e.g., flight costs for birds). However, for small animals that will perform behaviors in small, enclosed quarters, the technique can be highly successful. It has been used, for example, to measure the cost of burrowing in toads (Seymour, 1973) and gophers (Vleck, 1979); calling by frogs (Bucher et al., 1982; Taigen and Wells, 1985), katydids (Stevens and Josephson, 1977), and cicadas (MacNally and Young, 1981); pre-flight warmup in moths (Bartholomew et al., 1981); courtship and aggression in salamanders (Bennett and Houck, 1983); and construction of foam nests by frogs (Bucher et al., 1982; Ryan et al., 1983). In my opinion, this technique has been underutilized in studies of behavioral energetics. Its technical problems have already been solved and numerous, interesting problems remain to be investigated. These include, for example, the cost of singing by birds, territorial aggression in fish, prey consumption by snakes, and territorial displays by lizards, to name only a few.

3. Measuring the Energetics of Non-sustainable Behaviors

Intense activity demands rates of energy utilization above the capacity of the aerobic metabolic systems. Supplementary anaerobic metabolism provides for that energy delivery but results in endproduct accumulation and ultimate and often rapid fatigue. Anaerobic metabolic pathways vary in different taxa (Bennett, 1978; Hochachka and Somero, 1984). Lactic acid production, for instance, is restricted to chordates, crustaceans and some arachnids and annelids. Among molluscs, bivalves accumulate a variety of endproducts during anoxic exposure, including succinate and proprionate; cephalopods form octopine during activity. Insects do not use anaerobic metabolism, at least in their flight muscles.

Most measurements of anaerobic metabolism during activity have been undertaken on vertebrates, although a parallel series of observations is now being made on crustaceans (e.g., Burke, 1979; Full and Herreid, 1984; McMahon et al., 1979) and spiders (e.g., Prestwich, 1983). These have usually involved determination of lactic acid accumulation, lactic acid being the principal anaerobically-formed endproduct in all these animals. Depletion of high energy phosphate stores may also contribute to anaerobic energy production but this is more difficult to analyze and has not often been attempted: a significant depletion of phosphocreatine was found in tadpoles during burst swimming (Gatten et al., 1984) and changes in total high energy phosphate pools have been demonstrated by nuclear magnetic resonance in an active lizard (Smith and Schmidt, 1983). The latter technique, while potentially promising for in vivo metabolic measurements, currently has substantial difficulties in quantification of compound interconversions and in excessively long sampling times. Lactate, in contrast, is easily measured in either body fluids or tissue samples by spectrophotometric techniques. For use in the measurement of anaerobic energetics, it is best to analyze the accumulation of lactate in the entire body of the animal (Bennett and Licht, 1972). Due to the compartmentalization of lactate formation and elimination, sampling only one body compartment (e.g., blood) is definitely less desirable and involves many assumptions about lactate distribution. Measurements of total body lactate formation involve determination of the lactate content of a group of animals before a behavioral event and the content of another group that has engaged in the behavior. An alternate design might involve measurement of a series of post-active groups after different durations or intensities of behavior. The technique has two primary difficulties. First, pre- and post-active determinations cannot be done on the same individual animals. Consequently, careful attention to pre-active controls is necessary. Second, the method is restricted to animals of small body size (i.e., under 1 kg). For larger animals, blood sampling is an inferior but still feasible technique as long as the limitations on estimating total lactate formation are recognized.

The utilization of anaerobic metabolism under free-ranging field conditions can be determined by measuring lactate concentrations in post-active animals. Blood samples are sufficient if only a qualitative result is desired; total body content is better if the specification of the quantity of joules or high energy phosphate equivalents is required. Samples taken under field conditions must be preserved immediately on collection by freezing in liquid nitrogen or by homogenization in a protein-precipitating acid. Measurements of anaerobic metabolism under free-ranging conditions have not been made frequently. This is another area in which a considerable amount of work remains to be done. Significant lactate accumulation has been found in lizards during natural activity and staged territorial encounters in the field (Bennett et al., 1981; Pough and Andrews, 1985a), in frogs in a breeding chorus (Pough and Gatten, 1984; but see also Ryan et al., 1983 and Taigen and Wells, 1985), in crocodiles during escape behavior (Bennett et al., 1985), in sea turtles during egg laying (Jackson and Prange, 1979), and in hibernating turtles (Gatten, 1981). Many natural behaviors thus appear to entail some anaerobic metabolism. It does not appear to be a system that is restricted for use only under desperate circumstances but rather one that is frequently used as a supplement to aerobic energy provision. However, the amount of lactate formed is usually low in comparison with the animal's capacity for lactate formation. Animals appear to structure their behavior to avoid extensive anaerobiosis with its consequent fatigue. Surprisingly, the circumstance in which anaerobic metabolism does appear to be avoided by free-ranging animals is that in which it was first described under laboratory conditions: diving. Most freely-diving animals undertake relatively short dives that can be supported by aerobic metabolism, reserving anaerobiosis only for extended dives or underwater emergencies (Kooyman et al., 1981; Seymour, 1982).

The use of anaerobic metabolism during a behavioral sequence can also be determined in the laboratory. This can be done under unrestricted conditions allowing free behavior. This approach has been used to measure the anaerobic energetics of a predator-prey encounter between garter snakes and salamanders (Feder and Arnold, 1982). Both animals have significant amounts of lactic acid formation and the salamanders approach fatiguing levels during their escape attempts. Newborn garter snakes also use their full anaerobic capacity during escape attempts (Arnold and Bennett, 1984). Measurements of lactate formation and oxygen consumption may be made simultaneously on animals active in closed chambers (as described previously). In this circumstance, the relative importance of aerobic and anaerobic energetic inputs can be measured simultaneously and the total cost of the behavior determined. This approach has been used to analyze the cost of courtship and aggression in a salamander (Bennett and Houck, 1983), nest construction, including fertilization and oviposition, by frogs (Ryan et al., 1983), and feeding by a lizard (Pough and Andrews, 1985b). Anaerobic metabolism does occur under these circumstances, but its energetic contribution is very small in comparison with that of aerobic metabolism. Again, it seems to me that these approaches have much to contribute to our understanding of the energetics of and metabolic support for behavior. They should be more widely utilized than they are.

4. Conclusions

We cannot trust our intuitive sense of how expensive a behavior might be. Some activities are surprisingly costly. Who, for example, would have anticipated that a calling frog raises its metabolic rate 400 to 2400% above resting levels (Bucher et al., 1982; Taigen and Wells, 1985)? Conversely, other expenses may seem low. Daily levels of free-ranging energy utilization exceed standard or basal levels by only 100 to 200% in a variety of different animal groups (King, 1974; Nagy, 1982). We might have anticipated that the costs of natural behavior, thermoregulation, growth, etc., would be considerably higher than these values indicate. It is simply not acceptable to assign metabolic costs by sheer guesswork (e.g., Orians' [1961] supposition of the cost of bird song and its subsequent incorporation into energy budget models, such as that of Schartz and Zimmerman [1971]). Additionally, the source of metabolic support may be poorly anticipated. The avoidance of anaerobic metabolism by diving animals (Seymour, 1982) and its use by calling frogs (Pough and Gatten, 1984) are both surprising findings. The study of behavioral energetics must be an empirical exercise until we know considerably more than we presently do.

Our capacity to investigate behavioral energetics under natural conditions in the field is greatly handicapped by our inability to measure or estimate oxygen consumption over short time intervals. The anaerobic contribution can be determined but is of minor energetic significance. I currently see no way of making desired energetic measurements of behavior on free-ranging animals under field conditions. We must rely instead on extrapolations from measurements on aerobic and anaerobic contributions to different behaviors that can be undertaken under laboratory conditions. This situation is not ideal, but it is better than nothing.

Acknowledgments

Support for the author was provided by NSF Grant PCM81-02331. I thank G. A. Bartholomew and F. H. Pough for their helpful comments on the manuscript and Martin Feder, George Lauder, and NSF Grant BSR83-20671 for the invitation and support to attend the symposium.

References

Arnold, S. J., and A. F. Bennett. 1984. Behavioural variation in natural populations. III. Antipredator behaviour in the garter snake, *Thamnophis radix*. *Anim. Behav.* 32: 1108-1118.

Bartholomew, G. A., D. Vleck, and C. M. Vleck. 1981. Instantaneous measurements of oxygen consumption during pre-flight warm-up and post-flight cooling in sphingid and saturniid moths. *J. Exp. Biol.* 90: 17-32.

Bennett, A. F. 1978. Activity metabolism of the lower vertebrates. *Ann. Rev. Physiol.* 40: 444-469.

Bennett, A. F. 1980. The metabolic foundations of vertebrate behavior. *BioScience* 30: 452-456.

Bennett, A. F., T. T. Gleeson, and G. C. Gorman. 1981. Anaerobic metabolism in a lizard (*Anolis bonairensis*) under natural conditions. *Physiol. Zool.* 54: 237-241.

Bennett, A. F., and L. D. Houck. 1983. The energetic cost of courtship and aggression in a plethodontid salamander. *Ecology* 64: 979-983.

Bennett, A. F., R. B. Huey, and H. B. John-Alder. 1984. Physiological correlates of natural activity and locomotor capacity in two species of lacertid lizards. *J. Comp. Physiol.* 154: 113-118.

Bennett, A. F., and P. Licht. 1972. Anaerobic metabolism during activity in lizards. *J. Comp. Physiol.* 81: 277-288.

Bennett, A. F., and J. A. Ruben. 1979. Endothermy and activity in vertebrates. *Science* 206: 649-654.

Bennett, A. F., R. S. Seymour, D. F. Bradford, and G. J. W. Webb. 1985. Mass-dependence of anaerobic metabolism and acid-base disturbance during activity in the salt-water crocodile, *Crocodylus porosus*. *J. Exp. Biol.*, in press.

Bucher, T. L., M. J. Ryan, and G. A. Bartholomew. 1982. Oxygen consumption during resting, calling, and nest building in the frog *Physalaemus pustulosus*. *Physiol. Zool.* 55: 10-22.

Burke, E. M. 1979. Aerobic and anaerobic metabolism during activity and hypoxia in two species of intertidal crabs. *Biol. Bull.* 156: 157-168.

Feder, M. E., and S. J. Arnold. 1982. Anaerobic metabolism and behavior during predatory encounters between snakes (*Thamnophis elegans*) and salamanders (*Plethodon jordani*). *Oecologia* 53: 93-97.

Flint, E. N., and K. A. Nagy. 1984. Flight energetics of free-living sooty terns. *Auk* 101: 288-294.

Full, R. J., and C. F. Herreid II. 1984 Fiddler crab exercise: the energetic cost of running sideways. *J. Exp. Biol.* 109: 141-161.

Garland, T., Jr. 1982. Scaling maximal running speed and maximal aerobic speed to body mass in mammals and lizards. *Physiologist* 25: 338.

Garland, T., Jr. 1983. Scaling the ecological cost of transport to body mass in terrestrial mammals. *Am. Nat.* 121: 571-587.

Gatten, R. E., Jr. 1981. Anaerobic metabolism in freely diving painted turtles (*Chrysemys picta*). *J. Exp. Zool.* 216: 377-385.

Gatten, R. E., Jr. 1985. The uses of anaerobiosis by amphibians and reptiles. *Am. Zool.*, in press.

Gatten, R. E., Jr., J. P. Caldwell, and M. E. Stockard. 1984. Anaerobic metabolism during intense swimming by anuran larvae. *Herpetologica* 40: 164-169.

Hargrove, J. L., and J. A. Gessaman. 1973. An evaluation of respiratory rate as an indirect monitor of free-living metabolism. In *Ecological energetics of homeotherms*, ed. J. A. Gessaman, pp. 77-85. Logan: Utah State Univ. Press.

Hochachka, P. W., and G. N. Somero. 1984. *Biochemical adaptation.* Princeton: Princeton Univ. Press.

Jackson, D. C., and H. D. Prange. 1979. Ventilation and gas exchange during rest and exercise in adult green sea turtles. *J. Comp. Physiol.* 134: 315-319.

Johnson, S. F., and J. A. Gessaman. 1973. An evaluation of heart rate as an indirect monitor of free-living energy metabolism. In *Ecological energetics of homeotherms*, ed. J. A. Gessaman, pp. 44-54. Logan: Utah State Univ. Press.

King, J. R. 1974. Seasonal allocation of time and energy resources in birds. In *Avian energetics*, ed. R. A. Paynter, pp. 4-70. Cambridge, Mass.: Nuttall Ornithological Club.

Kooyman, G. L., M. A. Castellini, and R. W. Davis. 1981. Physiology of diving in marine mammals. *Ann. Rev. Physiol.* 43: 343-356.

LeFevre, E. A. 1964. The use of D_2O^{18} for measurement of energy metabolism in *Columba livia* at rest and in flight. *Auk* 81: 403-416.

MacNally, R., and D. Young. 1981. Song energetics of the bladder cicada, *Cystosoma sandersii*. *J. Exp. Biol.* 90: 185-196.

McMahon, B. R., D. G. McDonald, and C. M. Wood. 1979. Ventilation, oxygen uptake and haemoglobin oxygen transport in the Dungeness crab *Cancer magister*. *J. Exp. Biol.* 80: 271-285.

Nagy, K. A. 1975. Water and energy budgets of free-living animals. Measurement using isotopically labeled water. In *Environmental physiology of desert organisms*, ed. N. F. Hadley, pp. 227-245. Stroudsburg, Penn.: Dowden, Hutchison, and Ross, Inc.

Nagy, K. A. 1982. Energy requirements of free-living iguanid lizards. In *Iguanas of the world: their behavior, ecology, and conservation*, ed. G. M. Burghardt and A. S. Rand, pp. 49-59. Park Ridge, N.J.: Noyes Publications.

Orians, G. H. 1961. The ecology of blackbird (*Agelaius*) social systems. *Ecol. Monogr.* 31: 285-312.

Pough, F. H. 1983. Amphibians and reptiles as low-energy systems. In *Behavioral energetics*, ed. W. P. Aspey and S. Lustick, pp. 141-188. Columbus: Ohio State Univ. Press.

Pough, F. H., and R. M. Andrews. 1985a. Use of anaerobic metabolism by free-ranging lizards. *Physiol. Zool.* 58: 205-213.

Pough, F. H., and R. M. Andrews. 1985b. Energy costs of subduing and swallowing prey for a lizard. *Ecology*, in press.

Pough, F. H., and R. E. Gatten, Jr., 1984. The use of anaerobic metabolism by frogs in a breeding chorus. *Comp. Biochem. Physiol.* 78A: 337-340.

Prestwich, K. N. 1983. The roles of aerobic and anaerobic metabolism in active spiders. *Physiol. Zool.* 56: 122-132.

Ryan, M. J., G. A. Bartholomew, and A. S. Rand. 1983. Energetics of reproduction in a neotropical frog *Physalaemus pustulosus*. *Ecology* 64: 1456-1462.

Schartz, R. L., and J. L. Zimmerman. 1971. The time and energy budget of the male dickcissel (*Spiza americana*). *Condor* 73: 65-76.

Schmidt-Nielsen, K. 1984. *Scaling: why is animal size so important?* New York: Cambridge Univ. Press.

Seymour, R. S. 1973. Physiological correlates of forced activity and burrowing in the spadefoot toad, *Scaphiopus hammondi*. *Copeia* 1973: 103-115.

Seymour, R. S. 1982. Physiological adaptations to aquatic life. In *Biology of the Reptilia*, Vol. 13, ed. C. Gans and F. H. Pough, pp. 1-51. New York: Academic Press.

Smith, E. N., and P. G. Schmidt. 1983. Effect of temperature and exercise on the phosphorus metabolites of the lizard *Anolis carolinensis*. *Fed. Proc.* 42: 469.

Stevens, E. D., and R. K. Josephson. 1977. Metabolic rate and body temperature in singing katydids. *Physiol. Zool.* 50: 31-42.

Taigen, T. L., and K. D. Wells. 1985. Energetics of vocalization by an anuran amphibian (*Hyla versicolor*). *J. Comp. Physiol.* 155: 163-170.

Taylor, C. R., N. C. Heglund, and G. M. O. Maloiy. 1982. Energetics and mechanics of terrestrial locomotion. I. Metabolic energy consumption as a function of speed and body size in birds and mammals. *J. Exp. Biol.* 97: 1-21.

Townsend, C. R., and P. Calow. 1981. *Physiological ecology: an evolutionary approach to resource use.* Sunderland, Mass.: Sinauer.

Vleck, D. 1979. The energy cost of burrowing by the pocket gopher *Thomomys bottae. Physiol. Zool.* 52: 122-136.

6 A Comparative Approach to Field and Laboratory Studies in Evolutionary Biology

Raymond B. Huey and Albert F. Bennett

1. The Comparative Approach

1.1 Introduction

The comparative method is an important tool for analyzing interspecific patterns at all levels of biological organization. Versatility is one key to the method's success. Consider its diverse applications to issues that emerge from a simple observation that two species of predators differ in movement rate. A comparison of physiological capacities of the two species might pinpoint mechanistic bases underlying differences in movement rate and stamina, a comparison of prey types or of social systems might clarify the significance of movement rate in the ecology and behavior of predators, and a comparison of net energetic gains or of predation risk might hint at the historical pressures that led to the evolutionary divergence in movement rates.

Data in comparative studies are obviously varied and may be gathered from the field, the laboratory, or the literature. Nevertheless, comparative data typically share one characteristic: most are descriptive, not experimental. As a result, conclusions in comparative studies are usually based on static differences or correlations (e.g., stamina is correlated with oxygen transport capacity), not on dynamic responses (e.g., stamina is directly influenced by experimental manipulation of oxygen transport capacity).

In this paper, we evaluate the effectiveness of descriptive comparisons in answering mechanistic, ecological, and evolutionary questions. We illustrate our views with field and laboratory data on the evolutionary ecology of foraging mode in lacertid lizards from the Kalahari Desert of Africa.

Four main conclusions emerge from our analysis. First, descriptive comparisons are an efficient and effective way of revealing mechanistic, ecological, and evolutionary patterns in nature. Second, descriptive comparisons nevertheless have limited power in establishing cause and effect. Supplementary manipulative experiments will generally be required to substantiate the ecological processes underlying those patterns. Third, a phylogenetic perspective must accompany comparative studies in evolutionary ecology. A consideration of phylogeny not only guides the selection of species appropriate for comparison, but also suggests the likely direction of past evolutionary change. Fourth, because the analysis of the evolutionary processes that led to current patterns is accessible only by descriptive studies and not by experimental manipulations, definitive answers to evolutionary questions will necessarily be evasive.

1.2 The Problem: Foraging Mode in Kalahari Lacertid Lizards

Many studies of predation are stimulated by natural-history observations on animals in the field (Chapter 7). This one was as well. During a year-long study of the species diversity and ecology of Kalahari lizards, Pianka and Huey noted that different species of lacertid lizards differ conspicuously in foraging mode (Huey and Pianka, 1981; Pianka et al., 1979). Movement patterns of these lizards

TABLE 1: Movement Patterns of Adult Kalahari Lacertid Lizards

Species	Moves/min.	Percent Time Moving	Speed (km/h)
Sit-and-Wait			
EREMIAS LINEOOCELLATA	1.5	14.3	0.07
MEROLES SUBORBITALIS	1.8	13.5	0.06
Widely-Foraging			
E. LUGUBRIS	3.0	57.4	0.31
E. NAMAQUENSIS	2.8	53.5	0.28
NUCRAS TESSELLATA	2.9	50.2	0.37

Modified from Huey and Pianka (1981).

during summer are summarized in Table 1. Some species move more frequently, move a greater proportion of the time, and move farther per hour than did others. Overall movement rates appear dichotomized among these lizards, and interspecific differences are stable even after months in captivity (D. Kairns and P. J. Regal, personal communication). The very active lizards were described as "widely-foraging," whereas the more sedentary lizards were called "sit-and-wait" predators. Dunham (1983) has subsequently developed a more direct method of quantifying foraging mode. Similar differences in foraging mode have been described in many animal taxa (Eckhardt, 1979; Ruben, 1976; Schoener, 1971; Toft, 1981; Webb, 1984), but the differences are not always dichotomous or stable (Huey and Pianka, 1981; Pough, 1983; Taigen and Pough, 1983).

1.3 General Issues in Evolutionary Ecology

This differentiation in foraging mode provokes a series of questions concerning the significance of foraging mode at several levels of biological organization (see Table 2). Some of these questions reflect functional or ecological issues, whereas others reflect evolutionary ones. These questions are general and apply to many biological phenomena,

TABLE 2: Conceptual Issues in Ecological Physiology

Ecological Issues

1) What are the ecological correlates of the phenomenon?

2) What ecological factors maintain contemporary patterns?

Mechanistic Issues

3) What are the physiological, behavioral, and morphological bases of the patterns?

Evolutionary Issues

4) What was the direction of evolutionary change?

5) What selective factors promoted that change?

---

not just to predation. We wish to evaluate how well descriptive studies answer these fundamental questions. We will treat these questions in order, except questions 2 and 5, which will be discussed together.

1.4 The Importance of Phylogenetic Control in Comparative Studies

The questions in Table 2 can be addressed with respect to foraging mode by comparing various aspects of the biology of widely-foraging and of sit-and-wait predators. But the validity of answers to those questions depends fundamentally on the criteria used in selecting species for comparison (Clutton-Brock and Harvey, 1984; Harvey and Mace, 1982). All too often, availability or convenience guides the selection of species. Whenever possible, however, closeness of phylogenetic affinity should guide that selection. What attracted us to the Kalahari lacertids was not *just* that they differed in foraging mode, but that close relatives differed in foraging mode -- for such close relationships provided the phylogenetic control necessary for comparative studies.

Why should phylogenetic control be important? Patterns uncovered in comparative studies are the results of unplanned "natural" experiments that have occurred over evolutionary time. Any experiment, whether natural or manipulative, produces unambiguous patterns only when confounding variables such as temperature or size have been controlled. Natural experiments must also attempt to control for coincidental attributes that are a function only of the independent evolutionary histories of the species themselves. In general, the greater the phylogenetic distance between species being compared, the greater the chance that such attributes will confound the comparisons. Thus, the choice of closely related taxa should reduce the risk that coincidental differences will mask significant patterns or perhaps even induce artifactual ones in comparative studies (Jarman, 1982).

An example of how the lack of phylogenetic control can weaken comparative results emerges from an analysis of possible dietary correlates of foraging mode. Invoking some encounter-probability arguments (Gerritsen and Strickler, 1977), Huey and Pianka (1981) predicted that widely-foraging lizards are more likely than sit-and-wait lizards to encounter and eat prey, such as termites, that are patchily and unpredictably distributed (Wilson and Clark, 1977). Comparative dietary data from Africa, North America, and Australia supported their prediction (Huey and Pianka, 1981). However, the North American and Australian data are based on cross-family comparisons. In North

American deserts, for example, the teiid *Cnemidophorus tigris* is a widely-foraging predator, whereas iguanids are generally sit-and-wait predators. Unfortunately, teiids and iguanids differ in several characteristics, not just in movement rate. Specifically, many teiids have relatively well developed olfactory capabilities, whereas most iguanids are more visually oriented (Benes, 1969; Stebbins, 1948). Huey and Pianka (1981) pointed out that these differences should predispose a teiid to uncover more subterranean termites than an iguanid, even if both lizards foraged at the same rate. Although the sensory difference might well have co-evolved with foraging mode (Regal, 1978), it represents an uncontrolled variable that confounds an attempt to use descriptive data as an unambiguous test of a possible causal relationship between movement rate and diet (Huey and Pianka, 1981). Such problems may still occur even when close relatives are being compared (Chapter 2), but they should occur less frequently than when phylogeny is loosely controlled or ignored.

An alternative method that achieves phylogenetic control has been developed by S. J. Arnold (1983; Chapter 10). Individual (rather than interspecific) variation can be compared and correlated: for example, movement rate and incidence of termite eating can be compared among individuals. Arnold's approach is, however, useful only if individual differences are consistent and if the magnitude of the differences is greater than the measurement error.

2. Sequence of Investigation

2.1 Ecological Correlates

The first question we address is: What are the ecological correlates of foraging mode? This question is of interest because of numerous theoretical predictions concerning the relationship between foraging mode and various aspects of an animal's biology, such as diet, survivorship, and social systems. Using standard descriptive data, we can readily determine whether interspecific patterns correspond to predicted patterns.

Table 3 lists a sampling of the traits that we have examined, specific predictions that have been made about differences between widely-foraging and sit-and-wait lizards with respect to those traits, and whether our field data match those predictions. For example, the prediction that widely-foraging lizards should eat relatively more termites was substantiated by our dietary analyses (above). Where data are available, most theoretical predictions are supported by observed patterns in these Kalahari lizards. Moreover, parallel patterns have been found in other lizard groups (Huey and Pianka, 1981; Magnusson, personal communication), in tropical anurans (Toft, 1981; Taigen and Pough, 1983), and in tropical snakes (Henderson, 1984). Consequently, most patterns appear to be general.

How well do descriptive studies answer questions concerning ecological correlates of foraging mode? They are a very good and efficient way of uncovering patterns in nature, for determining their generality, and for establishing whether theory and observation conform. In general, the types of field data necessary for these evaluations are standard and relatively easy to obtain. Nevertheless, these studies are neither intended nor able to establish cause and effect. Because multiple theories can sometimes generate the same predictions, direct insights into cause and effect will generally require supplementary experimental studies that build on descriptive patterns (Paine, 1977). For example, abundances of termites could be manipulated or monitored, and the potentially differential effects on lizards with different foraging modes could then be observed. The

TABLE 3: Testing Theoretical Predictions Concerning Ecological Correlates of Foraging Mode in Kalahari Lacertid Lizards

Trait	Prediction	Reference for Prediction	Observations Consistent?
Diet	WF* eat patchily distributed prey (e.g., termites)	Huey and Pianka, 1981	Yes[1]
	WF encounter more prey per time	Gerritsen and Strickler, 1977	Yes[1,2]
Predation	WF more vulnerable to SW predators	Huey and Pianka, 1981	Yes[1]
Reproduction	WF have smaller relative clutch masses	Vitt and Congdon, 1978	Yes[1]
Home Range	WF have large home ranges	Schoener, 1971	Yes[3]
Social System	WF less territorial	Stamps, 1976	?
"Curiosity"	WF are more curious	Regal, 1978	Yes[4]
Sensory Mode	WF have acute olfactory senses	Regal, 1978	?

*WF = Widely-foraging.
[1]Tests from Huey and Pianka (1981)
[2]Tests from Nagy et al., 1984
[3]Huey, Bennett and Nagy (unpubl.)
[4]Kairns and Regal (unpubl.)
? = No relevant data available.

relevant experimental techniques are now well established and have
been used and advocated in various contexts by Dunham (1983),
Ferguson and Fox (1984), and Helfman (Chapter 9).

2.2 Mechanistic Bases

The second question we wish to address is: What are the
mechanistic (physiological, behavioral, or morphological) bases for
differences in foraging mode? For example, do widely-foraging and
sit-and-wait lizards differ in their capacity for speed and stamina? If so,
are these differences reflected in organ systems, organs, or their
constituent parts?

Here a problem emerges. Functional issues (see Chapters 2, 3, and
10) can be studied at a variety of levels, as shown in Table 4 -- so
which level is most appropriate? The answer depends on the precise

TABLE 4: Establishing Physiological Bases of Foraging Mode

Level	Prediction	Laboratory Observations Consistent?
Whole-Animal Performance	WF* have greater stamina	Yes[1]
	S&W+ have higher burst speeds	Yes[1]
Metabolism	WF have greater aerobic scope	Yes[2]
	S&W have greater anaerobic scope	Yes[2]
Tissue	WF have greater oxygen transport capacity (greater hematocrit and larger hearts)	Yes[2]
	WF muscles are more fatigue resistant	No[2]
	S&W muscles have faster contractile velocities	No[2]
Biochemical	WF muscles have greater activity of aerobic enzymes	No[2]
	S&W muscles have greater myo-fibrillar ATP-ase activity (index of maximal velocity of shortening)	No[2]

*WF = Widely-foraging.
+S&W = Sit-and-wait.
[1]Tests from Huey et al. (1984).
[2]Tests from Bennett et al. (1984).

question being asked, for the general question here actually has two
subsidiary parts.

(i) Does physiological capacity constrain foraging mode? For
example, do sit-and-wait lizards have insufficient stamina to maintain a
widely-foraging pace? This question is ecologically motivated.

(ii) Which functional aspects are responsible for an observed
difference? This question is mechanistically motivated. This distinction
between ecological and mechanistic questions in physiology is
sometimes overlooked, but is fundamental (Bartholomew, 1966; Feder,
1984; Huey and Stevenson, 1979). Indeed, this distinction serves as a
useful guide in selecting the level appropriate for a given study.

Clearly, ecological questions are best answered by examining
whole-animal performance as directly as possible. For example, to
establish whether stamina limits foraging mode, one can place lizards
on a treadmill moving at normal speeds and then directly compare
their capacities for stamina (measured as time until exhaustion, Huey et
al., 1984). Similarly, to examine the importance of relative speed in
predator-prey encounters, one can measure speed profiles of animals in
racetracks and then use these in computer simulations (Huey and
Hertz, 1984; Webb, 1976). An even more direct approach would be to
stage actual predator-prey encounters.

On the other hand, questions concerning the mechanistic bases of
an observed difference in whole-animal performance are necessarily
based on other levels. For example, one can examine various
components of oxygen transport and of metabolism (Table 4) in an
attempt to discover possible bases of observed differences in stamina
(Chapter 5; Bennett et al., 1984; Garland, 1984; Ruben, 1983). All too
often, however, the performance level is skipped, and lower-level data
are used to answer ecological questions. This is inappropriate. The
performance of the whole is not always predictable from the
performance of an isolated part (Bennett et al., 1984; Feder, 1984;
Chapter 2; but see Garland, 1984; Chapter 10). Jimmy the Greek
(personal communication) predicts athletic champions based on their
past performances, but not on their aerobic scopes or enzyme profiles.
Ecologists should do the same.

The risk of using lower-level data to predict whole-animal
performance is exemplified in our Kalahari studies, in which differences
in whole-animal performance are marked. In general, sit-and-wait
lizards had relatively great acceleration and speed but relatively low
stamina (Huey et al., 1984). In fact sit-and-wait lizards cannot maintain
a widely-foraging pace for even 15 minutes, suggesting that they are
physiologically constrained to being sit-and-wait lizards. But our studies

of lower levels of biological organization show differences only at a few levels, not at all (Table 4; Bennett et al., 1984). Clearly, an attempt to infer stamina and speed from contractile properties of muscles or from activities of key enzymes would have led to an incorrect prediction that these species are physiologically similar -- they are not.

Lower-level studies should thus be restricted to answering mechanistic questions (for example, which physiological systems account for observed differences in speed or stamina?). Even in this case the ability of lower-level experiments to answer such questions is sensitive to two problems:

(i) In approaching mechanistic questions, physiologists sometimes focus only on the physiology of the organisms and thus ignore potential interactions with behavior and morphology. This trend is understandable and productive, but it oversimplifies the integral nature of organismal performance (see Chapters 2, 3, and 4). Firing patterns of muscles are controlled by neurons, not by electrodes; and the power output of contracting muscles depends on their insertions on dynamical elements (limbs and jaws), not on force transducers. Physiology and morphology set interactive limits on what is possible, but behavior constrains what actually occurs (Hertz et al., 1982; Chapter 3). More integrative, multi-disciplinary approaches are required in studies of the mechanistic bases of animal performance.

(ii) Despite appearances to the contrary, many comparative "experiments" in physiological ecology (e.g., those listed in Table 3) are descriptive rather than experimental and consequently are subject to our earlier cautions concerning the limits of descriptive analyses in ecology. For example, the cardiovascular system was the only lower-level physiological system in which features were correlated with stamina in the Kalahari lacertids (Table 4). Although this strongly suggests a cause-and-effect relationship (see also Garland, 1984), causality can be established only by supplementary manipulative experiments (e.g., by studying the effect of controlled manipulations of oxygen carrying capacity on stamina). Physiology -- like ecology - - can be a descriptive science.

Finally, whole-animal studies not only provide a framework for mechanistic approaches, but they can also place a new perspective on theoretical issues. Available foraging models assume that stamina is unlimited and that trade-offs do not exist between speed and stamina (e.g., Norberg, 1977); yet neither assumption is generally true (Bennett, 1980; Huey et al., 1984; Webb, 1984; Chapters 3 and 5). An explicit consideration that physiological and morphological capacities are limited should enhance the realism of theoretical models.

2.3 Direction of Evolution

The fourth question (Table 2) concerns the direction of evolutionary change. For example, was widely-foraging or was sitting-and-waiting the evolutionarily derived foraging mode?

This question can be attacked by integrating descriptive field data on foraging mode of several different species with independently established phylogenies (Gittleman, 1981; Lauder, 1981; Ridley, 1983). Figure 1 shows two hypothetical phylogenies, with foraging mode indicated for each of the three species. In the phylogeny on the left, widely-foraging appears (based on parsimony) to be the derived foraging mode; but in the phylogeny on the right, sitting-and-waiting is probably derived (parsimony approaches are not without problems; see Felsenstein [1985]). We have superimposed field observations on foraging mode of several genera of lacertids on a tentative lacertid phylogeny, courtesy of E. N. Arnold (personal communication). Although foraging mode is unknown for many genera, the overall pattern suggests that the sit-and-wait foraging mode is probably derived in African lacertids, with one possible reversal. This pattern may hold for Hispaniolan tree snakes as well (Henderson, 1982).

This type of phylogenetic analysis has been underutilized as an evolutionary tool in most comparative analyses of predation and other problems (but see Feder, 1978; Harvey and Mace, 1982; Ridley, 1983; Arnold, 1984). This may be because most ecologists are not trained to appreciate a phylogenetic perspective or because established phylogenies are rarely available. In any case, phylogenies are an extremely useful tool for comparative studies: phylogenies not only

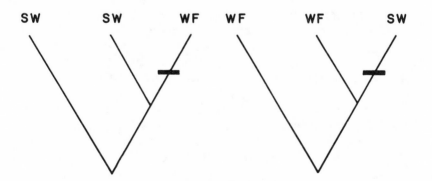

Fig. 1. Hypothetical phylogenies of three species, with foraging mode indicated for each (WF = widely-foraging, SW = sit-and-wait). The horizontal bar indicates an hypothesized evolutionary transition as established by parsimony.

guide the selection of close relatives for comparison (above), but also facilitate the interpretation of evolutionary patterns (Felsenstein, 1985).

2.4 Selective Factors

The final two questions (Table 2) concern selective factors that promote the maintenance and the origin of foraging mode. The question of maintenance is ecological, whereas that of origin is evolutionary.

These are clearly related issues, but a distinction between them is important -- factors that led to the origin of a trait are often different from those that currently maintain the trait. For example, venom of viperid snakes probably evolved initially for enhancing digestion, not for killing prey (Gans, 1978). Similarly, insect "wings" may have evolved initially for temperature regulation, not for flight (Kingsolver and Koehl, 1985).

What factors might be involved in the maintenance of different foraging modes? Theoretical models suggest the following predictions:

(i) If food is abundant, then widely-foraging lizards should have higher net energy gains than do sit-and-wait lizards; but if food is scarce, the reverse is true (Norberg, 1977).

(ii) On the other hand, sit-and-wait lizards, which should be relatively conspicuous to predators, should have lower rates of mortality (Gerritsen and Strickler, 1977).

(iii) If foraging mode influences diet, then differences in foraging mode may reduce interspecific competition (Pianka et al., 1979).

Do widely-foraging lizards have an energetic advantage? Net energy gain (estimated both with doubly labeled water and by short-term growth rates) is higher in widely-foraging lacertids (Nagy et al., 1984). Gross energy gain (size-corrected volumes of food in stomachs) is also higher in widely-foraging lizards (Huey and Pianka, 1981). Net energy gain is higher in widely-foraging teiid lizards than in sit-and-wait iguanids as well (Anderson and Karasov, 1981; Andrews, 1984). Thus, all available data suggest that widely-foraging lizards have higher net energy gains.

Do sit-and-wait lizards have a counteradvantage of reduced predation? Indirect morphological data (relative tail length, which is sometimes used as an index of predation intensity -- see Huey and Pianka [1981]) are supportive: widely-foraging lizards have relatively long tails. Moreover, sit-and-wait lacertids are eaten less frequently than expected (based on apparent densities) by horned adders (*Bitis caudalis*), a sit-and-wait predator (Huey and Pianka, 1981). Similarly, sit-and-wait lizards in North American deserts (H. W. Greene, personal communication) and in Hispaniola (Henderson, 1984) are eaten relatively infrequently by sit-and-wait snakes. Nevertheless, a direct analysis of the effects of foraging mode on predation rate awaits a capture-recapture study or other approaches.

Do differences in foraging mode reduce interspecific competition in food (Arnold, 1984; Pianka et al., 1979)? Foraging mode seems to influence diet, as discussed above. Nevertheless, experimentally establishing a link between foraging mode and competition may be difficult: the apparent stereotypy of foraging mode in these lizards, as shown in laboratory studies (Kairns and Regal, personal communication), suggests that removal experiments in nature might have little effect on foraging mode (see Schoener, 1974). But, for purposes of argument, assume that sit-and-wait lizards have lower mortality rates, that widely-foraging lizards always have higher net energy gains, and that differences in foraging mode reduce interspecific competition. Would these patterns -- alone or in combination -- be sufficient to establish an overall selective balance that promotes the coexistence of both foraging modes? Are they the only factors involved?

The question of selective balance is a difficult one in ecology (Paine, 1984). Descriptive studies can at best establish selective advantages and disadvantages associated with each foraging mode but will not establish whether those are of equal and opposite magnitude. Long-term analyses of density, survivorship, and reproductive output as functions of prey availability and of predator abundance may help, but definitive answers to questions concerning the coexistence of foraging modes will be difficult to obtain.

Unfortunately, question 5, which concerns factors that were responsible for the evolution of a trait in the first place, is even more difficult to answer. Recall that sitting-and-waiting is probably the derived foraging mode in lacertids and that sitting-and-waiting might evolve under three scenarios: if food is scarce, if predators are abundant, or if competition is reduced. But without knowing the precise environmental conditions under which evolution took place, and without knowing how abundant is "abundant," we are unable to determine the most reasonable of these scenarios. Moreover, even if those environmental conditions were known from paleontological evidence, we would not be able to use manipulative experiments to test our evaluations, as we can with contemporary issues.

In short, we will often be forestalled from answering questions of the evolutionary origin of complex traits. Much of what we need to know is lost in history, and we are constrained to working with descriptive rather than with manipulative tests. Perhaps the best we will be able to do is to say that the origin of a trait probably isn't related to factor A, but it might be related to factor B or factor C. Our potential to learn about historical phenomena seems inevitably limited. Given that evolutionary questions often provide the ultimate impetus for ecological and physiological research, this is frustrating.

3. Conclusion

We argue that descriptive, comparative studies are very useful in analyses of predation or of other phenomena. For many issues in evolutionary ecology, descriptive approaches are versatile, robust, and efficient: for example, they have uncovered a great deal about foraging mode of Kalahari lacertids. In addition, they set a solid framework for subsequent experimental analyses that more properly address processes involved in cause and effect. But the approach reaches its limits when extrapolating from patterns to processes or when extrapolating from the maintenance of traits to their historical origins.

Acknowledgments

We thank M. E. Feder and G. V. Lauder for the opportunity to participate in the symposium and this volume, anonymous reviewers for constructive comments on the manuscript, and E. N. Arnold for generously sharing his insights on the ecology and phylogeny of lacertid lizards. Research described here was funded primarily by NSF DEB 81-09667 to R. B. H., by NSF PCM 81-02331 to A. F. B., by the National Geographic Society to R. B. H., E. R. Pianka, and C. M. Cavalier, and by NSF GB-8727 to E. R. Pianka.

References

Anderson, R. A., and W. H. Karasov. 1981. Contrasts in energy intake and expenditure in sit-and-wait and widely foraging lizards. *Oecologia* 49: 67-72.

Andrews, R. A. 1984. Energetics of sit-and-wait and widely-searching lizard predators. In *Vertebrate ecology and systematics: a tribute to Henry S. Fitch*, ed. R. A. Siegel, L. E. Hunt, J. L. Knight, L. Malaret, and N. Zuschlag, pp. 137-144. Lawrence: Univ. of Kansas Museum of Natural History.

Arnold, E. N. 1984. Competition, evolutionary change and montane distributions. In *The evolving biosphere*, ed. P. L. Florey, pp. 217-228. Cambridge: Cambridge Univ. Press.

Arnold, S. J. 1983. Morphology, performance and fitness. *Am. Zool.* 23: 347-361.

Bartholomew, G. A. 1966. Interaction of physiology and behavior under natural conditions. In *The Galapagos*, ed. R. I. Bowman, pp. 39-45. Berkeley: Univ. of Calif. Press.

Benes, E. S. 1969. Behavioral evidence for color discrimination by the whiptail lizard, *Cnemidophorus tigris*. *Copeia* 1969: 707-722.

Bennett, A. F. 1980. The metabolic foundations of vertebrate behavior. *BioScience* 30: 452-456.

Bennett, A. F., R. B. Huey, and H. B. John-Alder. 1984. Physiological correlates of natural activity and locomotor capacity in two species of lacertid lizards. *J. Comp. Physiol.* 154: 113-118.

Clutton-Brock, T. H., and P. H. Harvey. 1984. Comparative approaches to investigating adaptation. In *Behavioural ecology: an evolutionary approach*, 2nd ed., ed. J. R. Krebs and N. B. Davies, pp. 7-29. Sunderland, Mass.: Sinauer.

Dunham, A. E. 1983. Realized niche overlap, resource abundance and intensity of interspecific competition. In *Lizard ecology: studies of a model organism*, ed. R. B. Huey, E. R. Pianka, and T. W. Schoener, pp. 261-280. Cambridge: Harvard Univ. Press.

Eckhardt, R. C. 1979. The adaptive syndromes of two guilds of insectivorous birds in the Colorado Rocky Mountains. *Ecol. Monogr.* 49: 129-149.

Feder, M. E. 1978. Environmental variability and thermal acclimation in neotropical and temperate zone salamanders. *Physiol. Zool.* 51: 7-16.

Feder, M. E. 1984. Consequences of air breathing for amphibian larvae. In *Respiration and metabolism of embryonic vertebrates*, ed. R. S. Seymour, pp. 71-86. Junk, The Hague.

Felsenstein, J. 1985. Phylogenies and the comparative method. *Am. Nat.* 125: 1-15.

Ferguson, G. W., and S. F. Fox. 1984. Annual variation of survival advantage of large juvenile side-blotched lizards, *Uta stansburiana*: its causes and evolutionary significance. *Evolution* 38: 342-349.

Gans, C. 1978. Reptilian venoms: some evolutionary considerations. In *Biology of the Reptilia*, Vol. 8, ed. C. Gans and K. A. Gans, pp. 1-42. New York: Academic Press.

Garland, T., Jr. 1984. Physiological correlates of locomotory performance in a lizard: an allometric approach. *Am. J. Physiol.* 247: R806-R815.

Gerritsen, J., and J. R. Strickler. 1977. Encounter probabilities and community structure in zooplankton: a mathematical model. *J. Fish. Res. Bd. Can.* 34: 73-82.

Gittleman, J. 1981. The phylogeny of parental care in fishes. *Anim. Behav.* 29: 936-941.

Harvey, P. H., and G. M. Mace. 1982. Comparisons between taxa and adaptive trends: problems of methodology. In *Current problems in sociobiology*, ed. King's College Sociobiology Group, pp. 343-362. Cambridge: Cambridge Univ. Press.

Henderson, R. W. 1982. Trophic relationships and foraging strategies of some New World tree snakes (*Leptophis, Oxybelis, Uromacer*). *Amphibia-Reptilia* 3: 71-80.

Henderson, R. W. 1984. The diets of Hispaniolan colubrid snakes. I. Introduction and prey genera. *Oecologia* 62: 234-239.

Hertz, P. E., R. B. Huey, and E. Nevo. 1982. Fight versus flight: body temperature influences defensive responses of lizards. *Anim. Behav.* 30: 676-679.

Huey, R. B., A. F. Bennett, H. B. John-Alder, and K. A. Nagy. 1984. Locomotor capacity and foraging behaviour of Kalahari lacertid lizards. *Anim. Behav.* 32: 41-50.

Huey, R. B., and P. E. Hertz. 1984. Effects of body size and slope on acceleration of a lizard (*Stellio (Agama) stellio*). *J. Exp. Biol.* 110: 113-123.

Huey, R. B., and E. R. Pianka. 1981. Ecological consequences of foraging mode. *Ecology* 62: 991-999.

Huey, R. B., and R. D. Stevenson. 1979. Integrating thermal physiology and ecology of ectotherms: a discussion of approaches. *Am. Zool.* 19: 357-366.

Jarman, P. J. 1982. Prospects for interspecific comparisons in sociobiology. In *Current problems in sociobiology*, ed. King's College Sociobiology Group, pp. 323-342. Cambridge: Cambridge Univ. Press.

Kingsolver, J. G., and M. A. R. Koehl. 1985. Aerodynamics, thermoregulation, and the evolution of insect wings: differential scaling and evolutionary change. *Evolution* 39: 488-504.

Lauder, G. V. 1981. Form and function: structural analysis in evolutionary morphology. *Paleobiology* 7: 430-442.

Nagy, K. A., R. B. Huey, and A. F. Bennett. 1984. Field energetics and foraging mode of Kalahari lacertid lizards. *Ecology* 65: 588-596.

Norberg, R. A. 1977. An ecological theory on foraging time and energetics and choice of optimal food-searching method. *J. Anim. Ecol.* 46: 511-529.

Paine, R. T. 1977. Controlled manipulations in the marine intertidal zone and their contributions to ecological theory. *The changing scenes in natural sciences, 1776-1976. Academy of Natural Sciences (Philadelphia) Special Publication* 12: 245-270.

Paine, R. T. 1984. Ecological determinism in the competition for space. *Ecology* 65: 1339-1348.

Pianka, E. R., R. B. Huey, and L. R. Lawlor. 1979. Niche segregation in desert lizards. In *Analysis of ecological systems*, ed. D. J. Horn, G. R. Stairs, and R. D. Mitchell, pp. 67-115. Columbus: Ohio State Univ. Press.

Pough, F. H. 1983. Amphibians and reptiles as low-energy systems. In *Behavioral energetics*, ed. W. P. Aspey and S. I. Lustick, pp. 141-188. Columbus: Ohio State Univ. Press.

Regal, P. J. 1978. Behavioral differences between reptiles and mammals: an analysis of activity and mental capabilities. In *Behavior and neurobiology of lizards*, ed. N. Greenberg and P. D. MacLean, pp. 183-202. Washington, D. C.: Dept. of Health, Education, and Welfare.

Ridley, M. 1983. *The explanation of organic diversity. The comparative method and adaptations for mating.* Oxford: Clarendon Press.

Ruben, J. A. 1976. Correlation of enzymatic activity, muscle myoglobin concentration and lung morphology with activity metabolism in snakes. *J. Exp. Zool.* 197: 313-320.

Ruben, J. A. 1983. Mineralized tissues and exercise physiology of snakes. *Am. Zool.* 23: 377-381.

Schoener, T. W. 1971. Theory of feeding strategies. *Ann. Rev. Ecol. Syst.* 2: 369-404.

Schoener, T. W. 1974. Resource partitioning in ecological communities. *Science* 185: 27-39.

Stamps, J. 1976. Social behavior and spacing patterns in lizards. In *Biology of the Reptilia*, Vol. 7, ed. C. Gans and D. W. Tinkle, pp. 265-334. New York: Academic Press.

Stebbins, R. C. 1948. Nasal structure in lizards with reference to olfaction and conditioning of the inspired air. *Am. J. Anat.* 83: 183-222.

Taigen, T. L., and F. H. Pough. 1983. Prey preference, foraging behavior, and metabolic characteristics of frogs. *Am. Nat.* 122: 509-520.

Toft, C. A. 1981. Feeding ecology of Panamanian litter anurans: patterns in diet and foraging mode. *J. Herpetol.* 15: 139-144.

Vitt, L. J., and J. D. Congdon. 1978. Body shape, reproductive effort, and relative clutch mass in lizards: resolution of a paradox. *Am. Nat.* 112: 595-608.

Webb, P. W. 1976. The effect of size on the fast-start performance of rainbow trout *Salmo gairdneri*, and a consideration of piscivorous predator-prey interactions. *J. Exp. Biol.* 65: 157-177.

Webb, P. W. 1984. Body form, locomotion and foraging in aquatic vertebrates. *Am. Zool.* 24: 107-120.

Wilson, D. S., and A. B. Clark. 1977. Above ground predator defence in the harvester termite, *Hodotermes mossambicus* (Hagen). *J. Ent. Soc. South Africa* 40: 271-281.

7 Natural History and Evolutionary Biology

Harry W. Greene

1. Introduction

Natural history has had profound and diverse influences on the life sciences. It was a cornerstone in the origins of modern evolutionary biology, and many of us began our careers as weekend naturalists, equipped only with curiosity and a spiral notebook. There are signs, however, that natural history has fallen from grace in the last decades of the Twentieth Century. It has even been called boring, muddled, uninspiring, old-fashioned, and of no value in itself. Nevertheless, most speakers in the symposium upon which this volume is based called for more "basic natural history data." This prompted me to pose several questions from the floor, and I am grateful to the editors for an opportunity to discuss these problems in the following essay.

Natural history focuses attention on organisms, on where they are and what they do in their environments, and is approximately synonymous with autecology plus descriptive ethology. It includes normal behavioral repertoires ("ethograms") and changes in external and internal states, insofar as these pertain to what organisms do. Natural history includes things like the chronology of reproductive events, clutch size, and ecdysis, but typically does not encompass fine details of hormonal cycles and other physiological phenomena.

I will argue that natural history provides an interpretive context for addressing both broader and narrower questions, sometimes in serendipitous and unexpected ways. It is the "idea and induction" part of "the" scientific method, the essential prelude to formulating hypotheses as well as the raw material for testing them. In all of these attributes, good natural history is exactly analogous to descriptive anatomy and alpha systematics. This is not a novel viewpoint (e.g., Schmidt, 1946; Wilson, 1973), but it deserves further emphasis and explication. I will draw examples that pertain to the evolutionary morphology and foraging ecology of lower vertebrates, but the implications extend to other organisms and topics. Strong et al. (1984) make a cogent plea for the role of natural history in formulating community studies.

2. What's Wrong with Natural History?

The reasons for recent disinterest in and condescension toward natural history perhaps include strict adherence to Popperian concepts of what constitutes science, without regard for the origins of theory; the widespread appeal of reductionism and a certain "technophilia" that have accompanied the rise of molecular biology (cf. Platt, 1964); and powerful, institutionalized pressures to deliver fast results (Tinkle, 1979). Negative attitudes undoubtedly also reflect the fact that natural history sometimes does fall short, in that the information given is so incomplete as to be of little value for broader purposes. This is true of many accounts of snake diets (e.g., Greene and Oliver, 1965), which consist of only taxonomic lists of prey, with no mention of relative prey size, sex, and other variables. As a result, such data are of little value in the recent surge of emphasis on the feeding biology of these animals (Pough, 1983).

That natural history is viewed as old fashioned and of limited value, even by some evolutionary biologists, is in contrast to attitudes toward anatomy. Old, beautifully illustrated, descriptive accounts of morphology continue to be used frequently and even held in reverence by some modern workers. Perhaps a key difference is that structure is usually the most salient and constant aspect of an animal's phenotype. Thus, regardless of one's theoretical orientation (Wake, 1982), the immediate aim for comparative anatomy always has been straightforward: describe carefully and fully everything that can be seen. This has been accomplished by dedicated, skillful workers, and it has benefited from an aesthetic appeal as well. Natural history has not sustained a comparable aura among evolutionary biologists, and much of it has been done anecdotally, seemingly as an afterthought and under the naive assumption that no special training, perspective, or effort was necessary.

Another reason for the differential treatment of anatomy and natural history might be that functional and evolutionary morphologists are increasingly disposed toward an approach to the study of structure that is both holistic and truly comparative (Gans, 1975; Lauder, 1982; Wake, 1982; Chapter 2), thereby enhancing the potential value of extensive, strictly descriptive work. By contrast, conceptual advances in the "outdoor branches" of organismal and population biology seem to have been accompanied by a narrowing of focus. Ecologists and ethologists increasingly seem to concentrate on studying clutch size, mating system, or some other restricted subset of an animal's biology, and to ignore or treat naively holistic and historical considerations (e.g., Clutton-Brock and Harvey, 1984; there are notable exceptions, e.g., Huey and Pianka, 1981; Janzen, 1983; Chapter 6).

Peters' (1980) eloquent and thought-provoking essay presented a sharp dichotomy between natural history and ecology that epitomizes the problems addressed herein. He portrayed natural history as art, a "contemplative and reflective activity," sometimes deeply satisfying but always of value solely to the individual observer. For Peters (1982), science is prediction; only as such is it worth doing as a professional activity, and only as such can it grapple with the pressing environmental issues of the day. Or, as a colleague once told me, "If you're not doing an experiment every moment, you're wasting time."

I have some sympathy for Peters' skepticism, but his position is incomplete. All predictive hypotheses inevitably arise inductively, either from logical consideration of observations or from some form of imagination. One might sit by the fire and wait until the ring structure of benzene appears as an image of a snake among the flames (as did Kekule [Dobzhansky et al., 1977]), but the combination of careful empiricism and a well-prepared mind seems like a good bet, too. High quality, publicly recorded natural history *is* data-in-waiting, simultaneously able to provoke theory and confront any number of previously unforeseen predictions.

Whatever the reasons for current attitudes, they have practical consequences. Natural history courses are conspicuously absent in some quarters and viewed as pointless anachronisms in others. There is no National Science Foundation panel that explicitly accomodates autecology and descriptive ethology, despite the fact that reliable, broadly representative natural history data will always be essential for testing important theories in ecology and systematic biology. As recently as 1983, the reviewer's instructions for ecological papers submitted to *Herpetologica* implied that manuscripts which simply described the life of a previously unstudied organism were not acceptable. Instead, publishable papers were to address broader issues, despite the fact that this journal routinely includes alpha taxonomic work.

3. What's Good about Natural History?

The concept of adaptation includes differential performance of different phenotypes under natural conditions (Chapters 2 and 10). Behavior thus serves as a functional couple between the structural features of an organism and its environment, and must be a component in the complete evolutionary analysis of form. My ongoing studies of feeding in California snakes (Greene, 1986) underscore this point, by demonstrating ways in which intrinsically identical prey can confront different predators with very different tasks.

The western whiptail lizard (*Cnemidophorus tigris*, Teiidae) forages diurnally by moving from bush to bush, and remains in a shallow, tight-fitting burrow at night. This common species is important in the diets of several desert snakes. The sidewinder (*Crotalus cerastes,* Viperidae) lacks rostral specializations for burrowing, possesses immobilizing and tissue destructive venom, and apparently catches relatively large *C. tigris* by striking them from under bushes during the day. The longnosed snake (*Rhinocheilus lecontei*, Colubridae) is a powerful constrictor, has a pointed rostral scale that presumably facilitates digging, is nocturnally active, and probably subdues sleeping whiptails in their burrows. The coachwhip (*Masticophis flagellum*, Colubridae) is a slender, diurnal snake with physiological specializations that permit unusually fast movements, and catches relatively small, active lizards by chasing them. Simply knowing that *C. tigris* is a major diet item for each of these species would mask the significant differences in how each species interacts with its prey, and thereby confound the analysis of putative feeding adaptations.

An anecdote from recent field studies of large, tropical vipers also illustrates a way in which simple observations can yield unexpected rewards. For several weeks in the summer of 1984, I watched daily a gravid female fer de lance (*Bothrops asper*) at Finca La Selva, in northeastern Costa Rica. This 2.2 kg animal confined her activities to an area of a few square meters in a treefall gap, basking in the open for several hours every day and resting at night in a tight coil among nearby fallen branches. She did not feed during the time that I watched her, and no standing water was available in the immediate environment. However, the integument of this species has a peculiar, beautiful texture (the local name, *terciopelo*, means velvet) and rain water always formed large, standing beads on her back. On several occasions the snake inclined her head against her coils and drank water droplets, once for over ten minutes. It might well be that water-holding properties constitute a major biological role for the skin in some tropical snakes, as is the case for certain desert lizards (Gans et al., 1982, for *Moloch horridus*) and snakes (Louw, 1972, for *Bitis peringueyi*). However, in the absence of field observations, a laboratory morphologist asking evolutionary questions about reptile skin in moist habitats probably could not have even considered this possibility (cf. Smith et al., 1982).

Natural history also sets realistic boundaries on theory. We don't need theory that applies to things that don't occur, but we do need to know what a global theory must encompass or how to restrict and subdivide it. Most foraging theory has been predicated on the assumption that organisms have "a fairly clear statistical expectation of the resources [they] will come upon" (page 59 in MacArthur, 1972), and that search and handling costs are sufficiently similar that they can be manipulated relative to each other. Given this fact, it probably is not surprising that many successful studies, in terms of data matching theoretical predictions, have come from predators feeding very frequently on abundant, small prey (cf. Krebs et al., 1983).

Other studies on tropical snakes at La Selva suggest that if a truly general foraging theory is possible, it must encompass a greater range of possibilities than previously considered. By using radiotelemetry, I estimated foraging rates in the bushmaster, *Lachesis muta*, a 2-4 kg, 2-3 m long viper (Greene and Santana, 1983). A female slept during the day and hunted in an alert posture every night. She used three sites for 3-15 days and traveled a total of ca. 50 meters in a 35 day period. On the 15th night at the third site (24th day of observation), she caught a rodent weighing at least 50% of her body weight. Judging from observations on captive vipers, immobilization and ingestion would have required no more than a few dozen minutes. The snake then rested for nine days before changing sites. More limited observations of other bushmasters were similar, and also indicated that hunting sites were frequently within one meter of a palm tree (*Welfia georgii*), the fruits of which are an important food for the rodents upon which *Lachesis* feeds. Using our movement and diet data, some reasonable assumptions, and standard metabolic equations, C. R. Peterson (personal communication) has calculated that an adult bushmaster needs only approximately six typical meals per year to support the energetic costs of maintenance and foraging movements.

The feeding dynamics of these large snakes present a stark contrast to those of small to medium size endotherms, animals that have large daily energy requirements (cf. Carpenter et al., 1983; Congdon and Tinkle, 1982), feed frequently, and make foraging decisions on a scale of seconds or minutes (Krebs et al., 1983). If bushmasters make decisions at all, they might do so on a scale of days or weeks! Given the apparently large disparity between search and handling costs in snakes (Godley, 1981; Greene, 1984), a bushmaster should probably try for any prey that it encounters, within certain functionally constrained limits (e.g., several species of rodents within a broad size range, but not birds).

We still do not know enough about the array of existing feeding biologies to state whether a continuum exists between hummingbirds and bushmasters, or if these animals represent two or more clusters of phenomena. Although textbooks present generalizations about foraging styles in diverse animals, there are actually exceedingly few careful, direct observations of hunting and its consequences in wild animals (cf. Carpenter et al., 1983; Table 14 in Curio, 1976; Greene, 1982). This is particularly true for species other than large mammals and birds in open habitats, although there are impressive exceptions (e.g., Pianka, 1982, for varanid lizards; Spencer and Zielenski, 1983, for pine martens).

4. What is the Future of Natural History?

The above examples present a dilemma. Thorough, satisfying answers to important questions in evolutionary biology will ultimately require detailed, autecological inventories for a wide spectrum of organisms. But who will do this work, and who will pay for it? My impression is that descriptive accounts accumulate usually as anecdotes or as side effects of long-term studies on more narrow topics, themselves subject to attitudinal and economic constraints (Tinkle, 1979). I surmise that broad, detailed, explicitly organism-centered studies are relatively rare today, at least in part because they are not "trendy," they usually require long-term efforts, their relevance to evolutionary biology is not appreciated universally, and there are very few direct avenues of support for them.

Good natural history is a source of timeless, priceless information for the biological sciences. It inspires theory as well as provides crucial data for answers to comprehensive, synthetic problems in ecology, ethology, evolution, and conservation biology. Despite this fact, natural history costs relatively little compared to the resultant benefits. It is too important to be left only to chance observations, unprepared minds, and ancillary benefits of other studies.

Natural history should be encouraged to flourish, and I will close with some suggestions:

(i) Theoreticians and laboratory workers should cultivate a respect for and communication with natural historians. The hubris and obscure, in-group jargon that characterize some literature are only self-serving, and discourage others from contributing to the empirical refinement of theory.

(ii) Systematic biologists should recognize that comparative natural history is an essential component of the complete study of morphological evolution. In the wake of recent criticism of "naive adaptationism" (e.g., Gould and Lewontin, 1979), it is important to stress that wild organisms *do* have surroundings. To ignore this fact as a matter of approach is to pursue, at least tacitly, an historical theory that is devoid of environmental context.

(iii) Journals that focus on particular groups of organisms should encourage the publication of high quality natural history papers, as an investment in the future testing of theories. Specialty journals should consider the example of the *Journal of Ecology* and publish regularly a small number of excellent, autecological studies.

(iv) Granting agencies should face up to the high, intrinsic, lasting value and low relative costs of good natural history studies. They should seek ways to insure the overt support of *high quality* work with explicit *potential* importance (rather than immediate answers) for general questions.

(v) Theoreticians have made tremendous contributions to evolutionary biology, often causing us to seek out things in nature that we never dreamed existed. It must be emphasized that I am not denying the importance of theoretical and experimental work, that I am not making excuses for sloppy natural history, and I do not advocate that ill-conceived studies should receive public support. Natural historians should eschew "field boot chauvinism" (Colwell, 1983) and strive to be well informed on new techniques and theory. For field observations to be useful maximally to others, and thereby have value beyond the immediate satisfaction of the observer, they must be sufficiently precise, accurate, and detailed to chronicle interesting phenomena (cf. Drummond, 1981). To do otherwise in today's world of shrinking habitats and spreading extinctions is a disservice to science and the organisms that we study.

Acknowledgments

I thank R. B. Huey, R. F. Inger, C. A. Luke, S. Naeem, E. R. Pianka, E. D. Pierson, F. A. Pitelka, D. B. Wake, and M. H. Wake for their comments on the manuscript. The Organization for Tropical Studies, D. A. Clark, and D. B. Clark facilitated the work at La Selva, which was done in collaboration with M. A. Santana and supported by the World Wildlife Fund--U. S., the Annie M. Alexander Fund of the Museum of Vertebrate Zoology, and out of pocket. The studies of California snakes have been supported by the National Science Foundation (BSR 83-00346).

References

Carpenter, F. L., D. C. Paton, and M. A. Hixon. 1983. Weight gain and adjustment of feeding territory size in migrant hummingbirds. *Proc. Natl. Acad. Sci. USA* 80: 7259-7263.

Clutton-Brock, T. H., and P. H. Harvey. 1984. Comparative approaches to investigating adaptation. In *Behavioural ecology: an evolutionary approach*, 2nd ed., ed. J. R. Krebs and N. B. Davies, pp. 7-29. Sunderland, Mass.: Sinauer.

Colwell, R. K. 1983. Review of *Theoretical ecology: principles and applications*, ed. R. M. May. *Auk* 100: 261-262.

Congdon, J. D., and D. W. Tinkle. 1982. Energy expenditure in free-ranging sagebrush lizards (*Sceloporus graciosus*). *Can. J. Zool.* 60: 1412-1416.

Curio, E. 1976. *The ethology of predation*. Berlin: Springer-Verlag.

Dobzhansky, T., F. J. Ayala, G. L. Stebbins, and J. W. Valentine. 1977. *Evolution*. San Francisco: W. H. Freeman Co.

Drummond, H. 1981. The nature and description of behavior patterns. In *Perspectives in ethology*, vol. 4, ed. P. P. G. Bateson and P. H. Klopfer, pp. 1-33. New York: Plenum Press.

Gans, C. 1975. *Biomechanics, an approach to vertebrate biology*. Philadelphia: J. P. Lippincott Co.

Gans, C., R. Merlin, and W. F. C. Blumer. 1982. The water-collecting mechanism of *Moloch horridus* re-examined. *Amphibia-Reptilia* 3: 57-64.

Godley, J. S. 1981. Foraging ecology of the striped swamp snake, *Regina alleni*, in southern Florida. *Ecol. Monogr.* 50: 411-436.

Gould, S. J., and R. C. Lewontin. 1978. The spandrels of San Marco and the Panglossian paradigm: a critique of the adaptationist programme. *Proc. Roy. Soc. Lond. B* 205: 581-598.

Greene, H. W. 1982. Dietary and phenotypic diversity in lizards: why are some organisms specialized? In *Environmental adaptation and evolution*, ed. D. Mossakowski and G. Roth, pp. 107-128. Stuttgart: Gustav Fischer Verlag.

Greene, H. W. 1984. Feeding behavior and diet of the eastern coral snake, *Micrurus fulvius*. In *Vertebrate ecology and systematics: a tribute to Henry S. Fitch*, ed. R. A. Seigel, L. E. Hunt, J. L. Knight, L. Maleret, and N. Zuschlag, pp. 147-162. Lawrence: Mus. Nat. Hist., Univ. Kansas.

Greene, H. W. 1986. A diet-based model of snake evolution. In preparation.

Greene, H. W., and G. V. Oliver, Jr. 1965. Notes on the natural history of the western massassauga. *Herpetologica* 21: 225-228.

Greene, H. W., and M. A. Santana. 1983. Field studies of hunting behavior by bushmasters. *Am. Zool.* 23: 897.

Huey, R. B., and E. R. Pianka. 1981. Ecological consequences of foraging mode. *Ecology* 62: 991-999.

Janzen, D. H., ed. 1983. *Costa Rican natural history*. Chicago: Univ. of Chicago Press.

Krebs, J. R., D. W. Stephens, and W. J. Sutherland. 1983. Perspectives in optimal foraging. In *Perspectives in ornithology*, ed. A. H. Brush and G. A. Clark, Jr., pp. 165-216. Cambridge: Cambridge Univ. Press.

Lauder, G. V. 1982. Historical biology and the problem of design. *J. Theor. Biol.* 97: 57-67.

Louw, G. N. 1972. The role of advective fog in the water economy of certain Namib Desert animals. *Symp. Zool. Soc. Lond.* 31: 297-314.

MacArthur, R. 1972. *Geographical ecology*. New York: Harper and Row.

Peters, R. H. 1980. From natural history to ecology. *Persp. Biol. Med.* 23: 191-203.

Pianka, E. R. 1982. Observations on the ecology of *Varanus* in the Great Victorian Desert. *West. Austr. Nat.* 15: 37-44.

Platt, J. R. 1964. Strong inference. *Science* 146: 347-353.

Pough, F. H. 1983. Feeding mechanisms, body size, and the ecology and evolution of snakes: introduction to the symposium. *Am. Zool.* 23: 339-342.

Schmidt, K. P. 1946. The new systematics, the new anatomy, and the new natural history. *Copeia* 1946: 57-63.

Smith, H. M., D. Duvall, B. M. Graves, R. E. Jones, and D. Chiszar. 1982. The function of squamate epidermatoglyphics. *Bull. Philadelphia Herp. Soc.* 30: 3-8.

Spencer, W. D., and W. J. Zielinski. 1983. Predatory behavior of pine martens. *J. Mammal.* 64: 715-717.

Strong, D. R., D. Simberloff, L. G. Abele, and A. B. Thistle, eds. 1984. *Ecological communities: conceptual issues and the evidence.* Princeton: Princeton Univ. Press.

Tinkle, D. W. 1979. Long-term field studies. *BioScience* 29: 717.

Wake, D. B. 1982. Functional and evolutionary morphology. *Persp. Biol. Med.* 25: 603-620.

Wilson, E. O. 1973. Review of *The Serengeti lion*, by G. B. Schaller. *Science* 179: 466-467.

8 Defense Against Predators

John A. Endler

1. Synopsis

Predation may be divided into five stages: detection, identification, approach, subjugation, and consumption. There are a variety of methods of reducing predation, and some mechanisms occur in all stages. The mechanisms are summarized, along with comments on discovering them, and testing predictions of their function and success. There are at least five main problems in detection and prediction: (1) specificity of sensory modes and predation conditions, (2) detection vs. function, (3) multiple causal pathways, (4) multiple defense mechanisms, and (5) multiple trait functions. There is clearly no "best" predator defense because many have the same functions and different mechanisms have varying success for different predators. Little is known about defense in non-visual modes, or about integration of different mechanisms.

2. Introduction

Predation is important in the life of most animals, and can be an important cause of natural selection in prey species. It is sufficiently important in ecology that there is a long-standing argument about whether predation, competition, or abiotic factors are the major organizing factors in ecological systems. Thus the study of anti-predator defenses has broad implications.

Although studied for about 150 years, work on anti-predator defense is still mostly in the descriptive stage. There have been few attempts to make and test hypotheses about the functions of defenses, and even fewer attempts at predicting the modes of defenses. This makes a summary of the strengths and weakness of the subject very difficult. I will first attempt to organize and summarize the major anti-predator defense mechanisms by reference to a stylized predation event (Section 3), discuss the methods and problems in their study (Sections 4 and 5), and finally, identify some general patterns and some of the major gaps in our understanding of the subject (Section 6).

I will make no assumptions about whether the defense mechanisms originated by natural selection ("adaptations"), or whether they serve an anti-predation function by chance ("preadaptations" [Simpson, 1944], "protoadaptations" [Gans, 1974], or "exaptations" [Gould and Vrba, 1982]). Since even the latter can be "improved" by natural selection, I will merely explore the functional significance of the defense mechanisms. In keeping with the aims of this book, no attempt will be made to review the literature, and the last section will contain unabashed speculation. Entrees into the literature can be found in Cott (1940), Curio (1976), Edmunds (1974), Endler (1978, 1984, 1985), and Harvey and Greenwood (1978).

3. An Outline of Anti-predation Mechanisms

A successful predation event involves at least five stages: *detection, identification, approach, subjugation,* and *consumption* (Table 1). Anti-predator defense mechanisms (hereafter called "defenses") function by interrupting this sequence and serve to reduce the probability that the process will go to completion (see also Fig. 1 in Chapter 3). Consequently, some defenses may occur during any stage of the sequence. Some defenses serve at more than one stage, so the five stages should not be thought of as rigid or mutually exclusive units, but rather as signposts in the predation sequence. Table 1 is not meant to be exhaustive, but serves only to indicate the diversity of defense mechanisms. The rest of this section elaborates the table and points out some little-known gaps in our knowledge.

3.1 Detection

A predator must first detect its prey as an object which is distinct from the background. Thus, the first stage of an anti-predator defense

TABLE 1: Stages in a Successful Predation Event and Methods of Defense by Prey (V) against Predators (P) at each Stage

1. DETECTION

   A. **Rarity** reduces the random encounter rate betwen P and V, and reduces the risk still further if P's exhibit frequency-or density-dependent behavior (see Curio, 1976).

   B. **Apparent Rarity**
      i. differences between P and V activity periods
      ii. hiding or inconspicuous resting places
      iii. polymorphism

   C. **Immobility** works for any sensory mode which detects motion

   D. **Crypsis**; reduces signal to noise ratio of V in P's sensory field
      i. general resemblance (Cott, 1940), eucrypsis, or crypsis proper (Endler, 1981)
      ii. special resemblance (Cott, 1940) or masquerade (Endler, 1981)

   E. **Confusion**
      i. polymorphism
      ii. movement between constrasting sensory backgrounds (Endler, 1978)
      iii. random or unpredictable movement (see Curio, 1976). May also shift P's attention to other objects or other prey.
      iv. random or unpredictable sensory effects between P and V
      v. by extreme abundance (see Curio, 1976)
         a. predator saturation
         b. schooling or other concerted behavior

   F. **Distance, Acuity, and Perception Effects** (see Endler, 1978)
      i. minimum distance for non-detection of any spot or pattern element
      ii. minimum distance for non-detection of the color of a spot
      iii. "private wavelengths" (see also Lythgoe, 1979 and Endler, 1983)

2. IDENTIFICATION

   A. **Masquerade** (see 1.D above)
   B. **Confusion** (see 1.E above)
   C. **Aposematism and Warning Color/Sound/Smell** (see Ford, 1975)
   D. **Mullerian Mimicry** (see Ford, 1975 and Curio, 1976)
   E. **Batesian Mimicry** (see Ford, 1975 and Curio, 1976)

3. APPROACH

   A. **Mode of Fleeing**
      i. speed
      ii. sprint to cover
   B. **Random or Unpredictable ("Protean") Behavior** (see Curio, 1976)
   C. **Rush for Cover or Other Predator-Inaccessable Area**
   D. **Startle** (see Edmunds, 1974)

4. SUBJUGATION

   A. **Strength to Escape**
   B. **Mechanical Methods**
      i. physical toughness
      ii. mucus or slime
      iii. loose skin or tail autotomy (see Vitt, 1983)
      iv. spines or other structures
   C. **Noxiousness**
      i. spines and prickles
      ii. jaws and claws (bite and scratch at P)
      iii. bad tastes, toxins, and stings
   D. **Lethality**

5. CONSUMPTION

   A. **Emetic**
   B. **Poisonous**
   C. **Lethal**

is an attempt to reduce the signal-to-noise ratio which predators perceive, and so minimize the probability of detection. This can be done in at least six ways (Table 1). Some of these methods serve in later stages as well. Most work has been done with visual signals, and there has been remarkable neglect of defenses operating in hearing, olfaction, chemoreception, lateral line, and electrical sensory modes. The following discussion elaborates some parts of Table 1.

Rarity and predator behavior (1.A, Table 1). Frequency- and density-dependent predation (apostatic selection) results if a predator differentially preys upon common species or forms. This increases the protective effects of rarity, because the attack probability declines faster than prey density. In addition, such a predator may completely switch to other more common prey species. This behavior has been demonstrated in a wide variety of predators using vision (see Curio, 1976; Endler, 1978). It has been demonstrated only once for olfaction (Soan and Clarke, 1973), and has hardly ever been investigated in non-visual sensory modes.

Apparent rarity and activity timing (1.B.i, Table 1). A prey species can appear to be rare to its predator if there are temporal differences in activity between predator and prey. A common example is nocturnality as a protection against diurnal predators (and vice versa?), and another is emergence early in the spring before predators become common (*Ambystoma* salamanders). Very brief activity times or seasons may also increase apparent rarity, preventing predators from specializing on such prey and evolving counter-defenses. See also Chapter 9.

Polymorphism and apparent rarity (1.B.iii). Polymorphism is the presence of two or more distinct forms ("morphs") in the same population of a single species (Ford, 1975). Examples include guppies (*Poecilia reticulata*) and *Eleutherodactylus* tree frogs. For species with apostatic predators, polymorphism reduces the overall risk per individual because there are two or more rarer forms rather than a single more common form. In addition, increased apparent rarity may cause an apostatic predator to switch to other prey species. (Unless we assume kin selection, polymorphisms probably arose for reasons other than increasing the effective rarity of a species; polymorphism is an "exaptation" for defense by rarity). Polymorphisms may also confuse predators. So far these ideas are poorly documented.

Crypsis (1.D) can work in any sensory mode, but it has been thoroughly studied only in vision (Endler, 1978, 1983, 1984). Even within a sensory mode, there are many different ways to be cryptic against the same sensory background, so some species are polymorphic as well as cryptic (Endler, 1978). Crypsis applies not only to the body of a prey, but also to its nests or egg clutches, though this is little studied. There are two main forms of crypsis, crypsis proper (eucrypsis or general resemblance) and masquerade (special resemblance). In eucrypsis the prey's color pattern resembles a random sample of the background, whereas in masquerade the prey resembles a specific object which is not normally eaten. Crypsis may be achieved by means of permanent color patterns or structures, as in *Phrynosoma* or *Moloch* lizards, or by color changes or structures which make the prey less conspicuous when needed, as in *Chamaeleo* and *Anolis* lizards. Crypsis may also be aided by special behavior which aids the cryptic effect (as in *Monocirrhus* and *Polycentrus* leaf fish), ensures the proper background (*Syngnathus* fish, *Eleutherodactylus* frogs) or proper background alignment (*Syngnathus*, *Aeoliscus*, and *Diademichthys* fish and *Thecadactylus* lizards), or even insures the best ambient light conditions for crypsis (guppies, Endler, in prep.). It is not known what favors one particular mode of crypsis over another, nor what favors it in a particular sensory mode.

Confusion by unpredictable sensory effects (1.E.iv). Pulses of chemicals by fishes and amphibians in streams might affect contrast for chemosensory predators. Certain modes of swimming in certain habitats might affect contrast for predators using lateral line systems. On land and in water, certain forms of movement may affect auditory contrast for a sound-detecting predator. Unfortunately, crypsis and confusion mechanisms have not been investigated for non-visual mechanisms, and confusion has been little studied even for visual predators.

3.2 Identification

Once a prey has been detected as an object which is distinct from the background, it must be identified as edible or not. Many of the mechanisms which delay detection will also delay or eliminate recognition as edible prey items (Table 1). Two mechanisms with this dual function are masquerade and confusion. For example, in masquerade an animal which rests on and closely resembles a dead leaf will delay detection as something different from the background (crypsis proper), and if recognized as a discrete object, will delay recognition as an object different from the leaves (masquerade). Little work has been done on identification outside of visual predation. As in section 3.1, the discussion below elaborates some parts of Table 1.

Aposematism and mimicry (2.C-E). These depend upon recognition as discrete objects, and identification (true or false) as something with noxious properties. Conspicuous coloration (aposematic) is frequently associated with noxious or dangerous prey, as in Dendrobatidae, Salamandridae, and some Elapidae, or with aggressiveness, as in some moray eels. Aposematic devices need not be limited to colors, as in the head frills of *Chlamydosaurus* lizards (which can give a predator-inhibiting bite), and the hoods of cobras. Of course, these signals may also function in intraspecific communication, but this dual function has not been studied. Aposematic displays may be accompanied by behavior which emphasizes the signal; this has actually been shown to increase the avoidance learning rate of potential predators of *Taricha* (Johnson and Brodie, 1975). Aposematic devices may not even be limited to visual signals. Possible examples are the bite-warning hissing of some snakes and lizards, the rattling of *Crotalus*, and the odors of *Elgaria* and some snakes. Although only *Crotalus* is dangerous, the other's painful bites may inhibit all but the most specialized predators.

Mullerian mimicry has the advantage of giving predators fewer warning signals to learn, making identification more rapid and effective (Ford, 1975). It is known in coral snakes (Green and McDiarmid, 1981), and may also occur in salamandrids. It would be interesting to know if Mullerian mimicry is more common in prey species with predators of intermediate or much intelligence. Batesian mimicry is similar, but yields an incorrect identification, at least from the predators' viewpoint. It has been demonstrated in two model-mimic pairs *Notophthalmus* -- *Pseudotriton* and *Plethodon jordani* -- *Desmognathus*, and is very likely in two more pairs *Taricha torosa* -- *Ensatina eschscholtzi xanthoptica* and *Notophthalmus* -- *Plethodon* (Brodie and Howard, 1973; Hensel and Brodie, 1976; Howard and Brodie, 1973; Huheey and Brandon, 1974; Pough, 1974; Tilley et al., 1982). Batesian and Mullerian mimicry are possible in other sensory modes, but this has not been explored. A possible example is the tail vibration in many colubrid snakes, which sounds (on suitable substrates) like the rattle of *Crotalus*.

3.3 Approach

An attack usually follows identification of an animal as a prey species, and it can be subdivided into two stages, the approach and subjugation of the prey. An approach often consists of a slow approach or stalk, followed by a swift closing with the prey (chase); defenses may occur at any stage (Table 1). As in the previous sections, the following discussion elaborates parts of Table 1.

Mode of fleeing (3.A). If a prey's mode of travel is very different from the predator's, an advantage may be gained; examples include gliding frogs, snakes and lizards, flying fish, water "walking" basilisks, and to a lesser extent amphibians and semi-aquatic reptiles. The killifish *Rivulus hartii* will escape aquatic and semi-aquatic predators by leaping out of water and moving remarkably rapidly over the forest floor. In all these cases, the predator's locomotion is not as efficient as the prey's. Prey can obtain an additional advantage if they can sense when a predator starts a stalk. They can then initiate fleeing before the predator is ready for the final chase. This is more efficient than waiting for the attack because a prey which initiates the chase will have an advantage in reaction time, reaction distance, a planned escape route, and sometimes even a physiological advantage. Bauwens and Thoen (1981), Heatwole (1968), and Vitt (1983) give interesting discussions of the factors which favor various combinations of crypsis and reaction distances. This would repay further study.

Startle behavior (3.D). It is possible that predators may be startled, confused, or disoriented by sudden changes in body shape or color patterns, as in the axillary color patches in some *Eleutherodactylus*, *Hyla*, and *Phylomedusa* frogs, the bright ventral colors of some frogs (*Bombina*, *Pseudophryne*), salamanders (salamandrids), and snakes (many colubrids), and flashes of otherwise folded fins in many fishes. Many of these serve also in intraspecific communication, but no attempts have been made to distinguish the different functions, and few have tried even to demonstrate them directly. Hissing and striking by reptiles, and gaping or spreading the gill covers and fins by fishes, may also startle predators. The "unken" display postures (concavely curved body raised off the substrate) of salamandrids and bombinids serve to draw attention to the startling structures, and make the other defenses more efficient (Brodie, 1977; Johnson and Brodie, 1975; Brodie et al., 1984). The twitching of autotomized lizard tails, as well as the autotomy itself, may serve as and enhance startle and confusion effects (Dial and Fitzpatrick, 1983). Schools and aggregations may aggravate startle as well as confusion effects, since it is then more difficult for a predator to aim its attack at any one prey item. This is aided by frequent monomorphism of species and life stages (tadpoles) which aggregate or school; numerous studies have shown that the odd members of such groups are more often taken by predator than the common forms (Curio, 1976). A systematic study of startle effects has not been made.

3.4 Subjugation

If the stalk and chase are successful, the predator must subdue and kill the prey, and there are at least four methods which lower the probability of death of captured prey (Table 1). The following are some elaborations to Table 1.

Strength and toughness (4.A, 4.B). These allow a prey to free itself from a predator's jaws or limbs, and would be most important when the prey is about the same size as the predator. A survey of the distribution of relative sizes, strengths, and toughnesses of predators and prey would be interesting. For example, why have turtles and loricariid, callichthyid, and doradid catfish evolved armor, whereas other catfish and other aquatic animals have not?

Spines (4.B.iv) serve at least four purposes: (a) they make a prey prickly and noxious to handle and eat (*Moloch,, Phrynosoma*, many percomorph fishes); (b) they increase the effective cross-sectional diameter of prey, reducing the number of gape-limited predators which can eat them (*Gasterosteus*, doradid catfish); (c) they may function as active weapons against predators. For example, the spines at the base of the tail in *Ctenosaura* (and perhaps other lizards) are used very effectively when chased head-first into a hole. A light touch by the predator (or a human hand) is immediately followed by a powerful and painful thrust of the spines against the stimulus, sandwiching it between the spiny tail and the wall of the hole. (d) Spines may serve as delivery devices for painful toxins or irritants, as in the dorsal spines of *Noturus* catfish and scorpaenid fishes, the caudal spines of rays, and the exposed sharp-pointed ribs of *Salamandra, Pleurodeles*, and *Tylotriton* (Brodie, 1977; Brodie et al., 1984; Nowak and Brodie, 1978). The presence of spines and other mechanical defenses may also serve a warning function to visual predators. This brings up the possibility of mimicry when the spines are not very protective, but appear dangerous (*Bufo typhonius, Ceratophrys*, and horned vipers?). Such spines could also aid in crypsis. Much more work needs to be done on this problem.

Lethality (4.D) is a defense against a predator in the last stages of an attack; examples are dendrobatid frogs, venomous snakes and some fishes. These merely represent extremes in noxiousness, or in some cases, the use of the prey's own predatory mechanisms as a defense. Although the lethal prey may escape, the dead predator will have no opportunity to learn to avoid the individual or species in the future, so it is difficult to see how lethality can evolve (Ford, 1975; Greene and McDiarmid, 1981). It is possible that, if closely related lethal prey live together, lethality may spread by kin selection in the prey (Harvey et al., 1982). If the prey's lethal effect lowers the local predator density, then lethality and avoidance may spread by kin selection in prey and predator. This seems unlikely, and certainly sublethality is a better defense than lethality. In addition, what may be lethal to one predator, may be sublethal to another, and lethality to the first predator may be an accidental byproduct of an evolved defense against the second predator. Of course, an even better defense is to prevent the attack stage entirely.

3.5 Consumption

If an attack is successful, the predator must still eat the prey. Distasteful species may be emetic to predators which swallow their prey whole or nearly undamaged, and this may allow escape in some cases. Emetic and mildly poisonous species result in their predators learning to avoid them, but if the prey is killed, this defense can only evolve through kin selection in the prey (Harvey et al., 1982). However, in some cases of Mullerian mimicry, where some of the models are lethal, predators may learn to avoid the warning signal if they by chance attack only the *less* dangerous species (Mertensian mimicry). It would be interesting to know if those predator species which separate killing and eating (e.g. shrikes and many cats) suffer from poisons.

3.6 Final Comments on Predation Sequences

It is clear that much could be done to organize our ideas on anti-predator defenses, and search for patterns interrelating defenses at various stages. Some obvious questions include: How many defenses have evolved through phylogenetic accidents or were restricted through developmental constraints? Is there a limit to the combination of defenses which any one species can evolve? Is a particular combination of defenses found in particular habitats, associated with particular predators, or particular predatory modes?

4. Testing Hypotheses about Anti-predator Defenses

To fully understand anti-predator defenses we need to go beyond the purely descriptive stage, and attempt to test predictions about the ecology and evolution of defenses. There are two approaches to testing the hypothesis that a given trait functions as a defense against predators: (i) tests of differential survival of individuals with and without the trait (or with it more and less developed), and (ii) predictions about the details of the defense. These are two of the major methods of demonstrating natural selection (for a detailed discussion, see Endler [1985]). They are outlined briefly below.

Differential survival. If a species varies for a trait which is thought to serve as an anti-predator defense, then we can investigate its usefulness as a defense by comparing the survival rates of groups of cohorts with the mechanism, and those with a less developed or absent version. If the mechanism has little variation, then it can be artificially modified, for example by changing color patterns by painting, removing or adding spines, and so forth. This approach can directly demonstrate that the mechanism increases survival (for details see Ford, 1975, Endler, 1985, and Chapter 10). A related approach is to work out the mechanics or chemistry of the mechanisms's function during predation, but that would only demonstrate that the structure inhibits predation, not that a modification of the existing mechanism would necessarily increase mortality. However, it could allow refinement of predictions about the fitness of modified defenses.

Details of the defense. If enough is known about the biology of the prey and its predators, it is possible to make detailed predictions about the mode and function of the defense, rather than simply cataloguing the mechanisms. For example, the shift from immobility to fleeing in lizards as a predator approaches is predictable on the basis of: (i) speed of the predator's approach, measured by the visual field solid angle per second, (ii) size of the predator and its danger to the lizard, (iii) temperature, which affects response latency and fleeing speed, (iv) degree of crypsis and the visual abilities of the predators, (v) previous experience with predators, (vi) sex and body weight, and (vii) foraging mode and physical habitat structure of prey and predators (Bauwens and Thoen, 1981; Heatwole, 1968; Vitt, 1983; see also Chapters 3 and 6). Similar predictions and tests of predictions could be made with other species. This approach assumes that genetic variation and natural selection are great enough so that the mechanism can evolve towards the predicted equilibrium configuration (see discussion in Endler, 1985). That in itself is also a testable assumption, though that test has rarely been made. This method cannot by itself demonstrate that the mechanism actually works, so must be used in conjunction with a differential survival test (Endler, 1985).

An ability to quantify the trait which is thought to be an anti-predator defense greatly increases the efficiency of the formulation and testing of predictions, as well as making more predictions possible. A good example is crypsis in the sense of general resemblance (eucrypsis). An animal's color pattern can be defined as cryptic if it resembles a random sample of the background at the time and place at which it is most vulnerable to predation. Animal and background color patterns can be regarded as mosaics of patches which vary in color, reflectance, size, and shape. Thus a cryptic color pattern is a random sample of the color, reflectance, size, and shape distributions of the background. From this it follows that a measure of crypsis is the similarity between the animal and background distributions, and a statistical test for crypsis is a goodness-of-fit test between the two distributions (Endler, 1978, 1984). Using Pietriwicz and Kamil's (1981) experimental protocol with blue jays and *Catocala* moths on tree trunks, Endler and Kamil (in preparation) found that this measure of crypsis is highly correlated with the time from presentation to attack (Fig. 1). Thus the measure is more than an arbitrary quantification of crypsis. Given the definition and measure of crypsis we can make a large series of predictions about many aspects and modes of crypsis

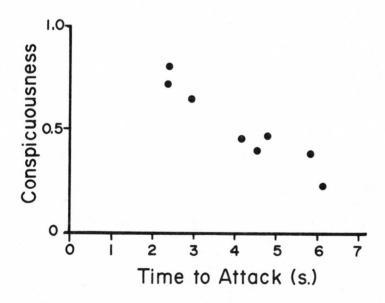

Fig. 1. Relationship between the time it takes blue jays to find and attack *Catocala* moths on various tree trunks and the moth's crypsis as measured by the similarity between the prey and background with respect to patch size and color frequency distributions.

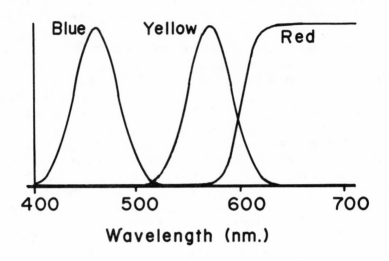

Fig. 2. The reflectance spectra of three colors, which are used in Figs. 3-5. The horizontal axis ranges from blue (400 nanometers) to red (700 nm).

(Endler, 1978), and many of them have been found to be correct (Endler, 1978, 1983, 1984). Similar detailed predictions could be made and tested for other sensory modes and other kinds of defense mechanisms, and this would be a productive avenue of research.

5. Problems

There are at least five kinds of problems which arise in studying anti-predator defenses: (1) specificity of sensory modes and conditions of predation, (2) detection vs.. function, (3) multiple causal pathways,

Fig. 3 (facing page). Method of estimating the brightness of a given color seen by a given animal under specified environmental conditions. Vertical axis: *a*, estimated spectral sensitivity of a guppy (*Poecilia reticulata*). *b*, guppy's sensitivity times ambient light spectrum in shallow water. *c*, absorbance spectrum of turbid fresh water. *d*, guppy's sensitivity times ambient light times turbidity times the blue spot reflectance in Fig. 2. See text for details.

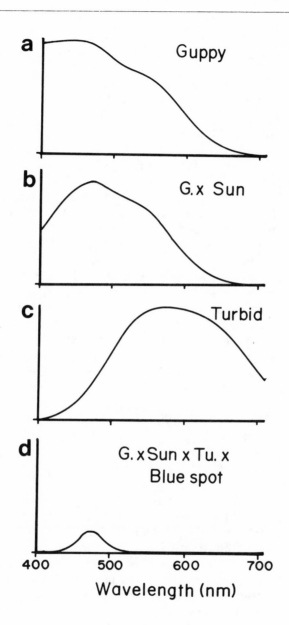

Wavelength (nm)

(4) multiple defense mechanisms or "strategies", and (5) multiple trait functions. As will become obvious in the discussion below, these can only be overcome with a detailed knowledge of the sensory biology and ecology of the predators and prey (see also Chapter 7).

5.1 Sensory Modes and Conditions

The detection and understanding of a particular anti-predator defense mechanism depends upon a thorough understanding of the sensory modes used by the predators and the biological and physical conditions affecting those senses during predation. Examples will be given for visual crypsis, but this discussion applies to other sensory modes and defense mechanisms.

One must know enough about the sensory processes so that one measures the correct range of parameters. For example, if a predator cannot distinguish red spots except by differential reflectance (brightness), then red should not be included in the comparison between the prey and background color frequency distributions. (However, the *reflectance* of the red spots should be included in the brightness frequency distribution.) Red and orange spots in two poeciliid fishes seem to be irrelevant to crypsis whenever the only or major predators are invertebrates (which are insensitive to red and orange), but these spots are very important to crypsis with respect to fish predators (Endler, 1978, 1983).

The environmental conditions during predation are very important and must be considered in any analysis. Lythgoe (1979) provides an exceptionally good discussion of the effects of environmental conditions on visibility of color patterns in terrestrial and aquatic habitats; analogous phenomena will be found in other sensory modes. As an example, consider three color reflectance spectra in Fig. 2. Consider a guppy (*Poecilia reticulata*) which has a spectral sensitivity approximately as shown in Fig. 3a. The curve is derived from single-cone spectral absorbance data (Levine and MacNichol, 1979; Levine, personal communication) used in Harosi's (1976) model of the relationship between cone absorbance and spectral sensitivity; it is therefore very approximate. If we multiply this by the incident light spectrum of sunlight in very shallow water we obtain the curve in Fig. 3b. If the water (fresh) is turbid, then the spectrum is further restricted as shown in Fig. 3c. Multiplying all factors together, including the reflectance curve of the blue spot of Fig. 2, we obtain a very low value for the brightness of the blue spot (Fig. 3d). Even though a guppy is very sensitive to blue, in turbid water there is little blue light, and therefore blue spots are not very bright in turbid water (for further discussion see Levine and MacNichol [1979] and Lythgoe [1979]). Undisturbed natural guppy populations are normally found in clear shallow water (Endler, 1978), where there is little attenuation of short wavelengths, so blue spots are very bright to their eyes, and are used in courtship. Note that this will be true even if there is no color discrimination. For example, two spots of different colors but equal reflectances will differ in perceived brightness under these conditions, even without perception as different colors. The relative brightness of the two spots will change with the retinal spectral sensitivity. (By analogy, consider what happens to a scene photographed in standard and infrared-sensitive black-and-white film.)

If conditions of measurement are different from those during predation, then misleading results are very likely (reviewed in Endler, 1978); conditions may change very rapidly. An example is the background in a forest during a sunny day, the same when the sun is obscured by a cloud, and at dawn and dusk. When the sun is unobscured, the background color pattern parameters in a forest consist of small patch size, large range (variance) of brightness, and a high color diversity, but at other times the mean patch size is much larger, the brightness variance (contrast) is small, and the color diversity is low (Endler, 1978). Some animals are active only when the sun is out, others only when the sun is obscured (or at dawn and dusk), others at all times but stay in shade, while still others stay in sun patches. This has a major effect in what the backgrounds can be, and in differences between measurement conditions and when predators are active (Endler, 1978). Measurements must be made under the same conditions as predation.

It is not sufficient to measure the environmental conditions during predation, but one must also take into account the sensory abilities of predators; animals' senses are often different than ours. A good example is shown in Figs. 4 and 5. Consider the three colors in Fig. 2 as they would be seen by various kinds of fishes under different conditions. Levine and MacNichol (1979) surveyed the absorption peaks of single cones from a wide variety of fishes and found that they fell into groups which corresponded roughly with habitats. For freshwater they divided the species into four groups: surface, midwater, crepuscular, and benthic, and these groups are increasingly red and decreasingly blue-sensitive. As was done for Fig. 3, a typical species was taken from each of the four groups and its data were put into Harosi's (1976) model. The resultant approximate spectral sensitivity curves were multiplied by the three color spectra in Fig. 2. The result was multiplied by the absorbance curves for various water and light conditions, which are summarized in Lythgoe (1979). The final result is the brightness of each of the three colors as it would appear to each of the four fish groups if they were to forage in four different conditions; see Figs. 4 and 5. Once again, it is not necessary to assume any color discrimination by these fish (although it may be present). In all water conditions, shorter wavelengths (bluer) are brighter for fish with more short-wavelength sensitivity, and vice versa; colors which are near the insensitive end of the sensitivity spectrum are not very bright. In clear shallow water all three colors are bright, after accounting for the spectral range of each retina type; for example, red spots are not very bright for the mostly blue-sensitive group I fish. As in Fig. 3, blue

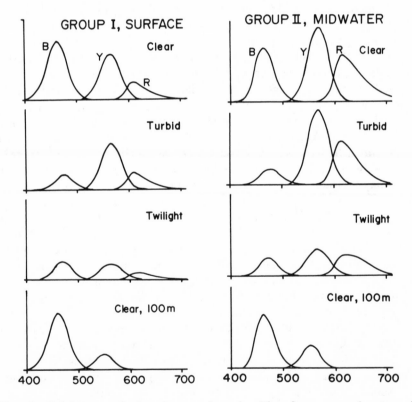

Fig. 4. Brightness of the three colors in Fig. 2 as seen by two fish retina/habitat classes under four conditions. For details see text. The guppy (Fig. 3) is a group 1 species.

spots are never very bright in turbid water, but yellow and red spots are. Note how groups II, III, and IV perceive reds more strongly (brighter) than group I; this is particularly marked in fish which are normally found in turbid waters. A similar effect is found for twilight. At 100 m depth there is so little long wavelength that only blues and yellows are seen at all, and they are seen most evenly (equally bright) by group IV fish. Clearly, the wrong assumptions about visual abilities and conditions of predation can yield quite false conclusions about relative and absolute brightness of various color pattern elements, and the crypsis or conspicuousness of prey species. These observations could easily apply to other sensory modes.

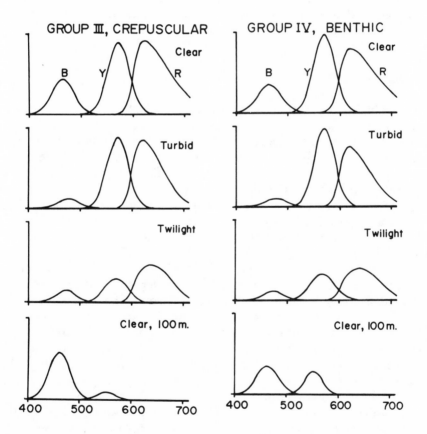

Fig. 5. Brightness of three colors for two more retina/habitat classes, as in Fig. 4.

5.2 Detection and Function

The detection of the advantage of a particular mechanism does not necessarily explain how it works. For example, we could do a capture-recapture study of sticklebacks which vary in the number of lateral plates. In some localities we would find that the number of lateral plates affects the probability of survival, and in fact Bell and Haglund (1978) showed that garter snakes (*Thamnophis*) do select for particular plate numbers. But it is not clear *why* they do so, and this cannot be solved without a detailed study of the functional anatomy of the plates and predator escape behavior. This is not to belittle such studies, only to point out that they are incomplete by themselves as efforts to understand anti-predator defense. Discussions of methods are found in Endler (1985) and Chapter 10, and the difference between detected advantage and function is discussed in Endler (1985).

5.3 Multiple Causal Pathways

The naive expectation about predictions holds that single causes have single effects, but this is quite often not the case. One effect may result from more than one cause and one cause may have more than one effect. Consequently, several mechanisms may yield the same predictions. For example, in natural populations of guppies, spot size declines from localities with few relatively innocuous predators to localities with many dangerous predators (Endler, 1978, 1983). This could have been predicted independently from three different causes: (i) The more dangerous predators have greater visual acuity; smaller spots are more likely to be below the predator's minimum detectable spot or color spot acuity limit. Even if acuity were equal among predator species, we might still expect smaller spots where there are more dangerous predators, because it would be more critically advantageous to be below the acuity limit in dangerous localities. (ii) More dangerous predators tend to be found downstream, and downstream localities have finer gravel and sand, while the relatively safe localities have larger gravel grain sizes. Background matching with respect to patch size predicts smaller spots in more dangerous localities. (iii) As male guppies grow older they slowly increase in size, and in some individuals the spot size actually increases with age. Since the mean life expectancy is shorter in areas of high predation, high predation guppies are on average younger, and therefore will have on average smaller spots than older low-predation guppies. Actually all three causes have been demonstrated by means of greenhouse and field transfer experiments (Endler, 1983, 1985). The third cause is an especially interesting one because it is a second-order rather than a direct effect of predation, but the net effect is the same (Endler, 1985).

5.4 Multiple Defense Mechanisms or "Strategies"

Antipredator defense can involve many different mechanisms (Table 1). There are often multiple "solutions" to the same defense "problem", and many different mechanisms work simultaneously. Therefore it is not surprising that multiple defense mechanisms are the rule.

Multiple mechanisms may make it difficult to interpret or even demonstrate the function of a supposed mechanism. For example, consider "disruptive coloration", a form of coloration which breaks up the body outline, making the prey more difficult to recognize (Cott, 1940; Edmunds, 1974). This is common in many fishes, amphibians, and reptiles. If most of the disruptive effect results from one white bar, then a test of the advantage of disruptive coloration might be to paint the color of the bar with some other color, painting another bar with the same paint in control animals. However, if the whole pattern is cryptic, then changing the white bar to another color may merely change the body from one cryptic pattern to another, and there will be no differences in survival between the painted and control animals. This has actually been found in experiments with insects (for example, Silberglied et al., 1980). Once again, a detailed knowledge of ecology and sensory biology of predation is essential.

5.5 Multiple Trait Functions

Traits used in defense may often be used in other functions, and this affects the accuracy of and tests of predictions. For example, color patterns may not only be cryptic or aposematic, but they can also serve in intraspecific communication and thermoregulation. If a given color pattern were thought to be used only in crypsis, but actually had other functions, then predictions based solely on the basis of crypsis would not be verified. The color patterns of guppies represent a balance between crypsis and sexual selection, and accurate predictions of the patterns are possible only when both factors are considered (Endler, 1983). By similar reasoning, the most cryptic color form of a polymorphic prey may not function well in thermal balance, but a less cryptic form might. The relationships between tail loss, tail length, crypsis, and fleeing are complex in lizards because the tail serves in foraging locomotion, balance, fat storage, and social position, as well as in predator escape (Dial and Fitzpatrick, 1983; Vitt, 1983). In addition, as Chapter 2 points out, a species can be a predator as well as a prey, so some mechanisms may have additional functions. A good example is crypsis or mimicry by predators which is used to hide from or mislead their prey (Cott, 1940; Edmunds, 1974; Endler, 1981). Multiple functions have been little studied, but are so common that they should not be ignored. This is an especially serious problem in ecological genetic studies, which usually concentrate on one trait and one function, or a few traits with one function each.

There may also be geographic variation in the chief functions of a trait, so it is dangerous to generalize from one locality to others (Endler, 1978, 1985). For example, *Pseudotriton ruber* is a mimic of the distasteful *Notophthalmus viridescens* only in moist areas where the latter are abundant, and cryptic in dry areas where the newt models are rare (Huheey and Brandon, 1974).

Even if a trait apparently has primarily only a single function, that function may work differently for different predators, or even for the same predator species under different conditions in different places. In that case, studying the wrong predator or conditions would be misleading. For example, a color pattern may exhibit the best background match for the most dangerous predator's visual abilities, yet be relatively conspicuous for other less dangerous predators. A study involving only the less dangerous predator would conclude that the species was not cryptic, and perhaps that it was invulnerable to predation. This will be a particularly difficult problem if two equally dangerous predators have differing visual abilities, or hunt under very different visual conditions. The problem may be mitigated if one predator is more dangerous, but has poorer vision, but this must be known. The effects of multiple predators with varying sensory abilities and predator conditions is completely unexplored and would repay further study.

6. General Observations, Predictions, and Speculations about Defenses

Since there are multiple functions of anti-predator defenses, and since more than one defense mechanism can be used at a time, there is no single "best" defense mechanism. Instead, suites of mechanisms are used. The combinations of mechanisms used may be the best combination for the particular community of predators, predator sensory capabilities, and environmental conditions during predation, or it may merely be the combination of traits which happen to have been used by the species' ancestors. This makes it difficult to test predictions about which defenses are used by species under known conditions. However, it is possible to understand the function of each mechanism from a thorough knowledge of the ecology and sensory biology of predation, and this can also allow successful prediction. There are major gaps in our knowledge of most of the defenses, especially where multiple functions and predators exist, but the gaps are especially great for non-visual sensory modes.

It is possible to make six tentative generalizations about anti-predator defenses at this time. (1) If something can be used as a defense mechanism, it will; almost any structure or feature of an organism can be used in defense. A remarkable example is the ribs pushing through the skin of salamandrids (Brodie et al., 1984). (2) There tends to be a shift from passive to active defense during a predation sequence. This may have arisen because as later stages are reached it becomes more and more important to actively prevent predation. (3) Defenses are more efficient if they stop the sequence as early as possible. This gives greater opportunity for sequence termination with minimal risk to the prey, and it also allows employment of a greater variety of defenses when others do not work. (4) Defenses tend to work later in the sequence for species which are distasteful compared to those that are not. It is difficult to know which is the cause and which is the effect of this phenomenon, but it has not been investigated. (5) Different sensory modes may be used at different stages of a predation event. For example vision and hearing are probably more effective during detection and identification, whereas tactile, chemosensory, lateral line and electric senses may be more efficient during the approach (but not in turbid or rapidly flowing water). (6) Phylogenetically closely related prey tend to have similar predators and also similar defenses. It is difficult to know if this is due to parallel evolution due to phylogenetic or developmental constraints, parallel or convergent evolution due to coevolution with the predators, or other common selective factors. Geographic variation in predator species correlated with geographic variation in defenses suggests selection rather than constraints (Endler, 1985). Unfortunately this is little studied and long term changes in predator distributions may destroy such patterns. Clearly much more needs to be done on all of these generalizations.

The following eight predictions are largely untested: (1) Certain suites of defense mechanisms may be found (a) in certain habitats, (b) associated with certain predator types, (c) associated with certain predator mechanisms, or (d) associated with certain predator sensory modes. Examples are (a) vertical bars in cryptic color patterns in animals which live in grassy places, and (b) sliminess and predators with long teeth (mollusk, fish, and amphibian prey). (2) Prey should be most cryptic, or have their best defenses, with respect to the sensory mode of the most dangerous predator. (3) Predators should use the most efficient sensory system for detecting and identifying prey. Since the most efficient sense depends upon phylogenetic constraints, the environmental conditions during predation, and the biophysics of the senses, and these change in geological time, different predators will use different sensory modes. As a result, a diversity of defenses will evolve in a species during its history. As a corollary, there may be geographic variation in defenses in species with large enough ranges. (4) Defenses should be directed at the predator with the best learning ability, even if it is not necessarily the most sensorily acute species. Predictions about defenses will be difficult in the absence of such information. (5) Because masquerade requires complex pattern recognition and eucrypsis does not, masquerade may be more common in species whose most dangerous predators are capable of complex pattern recognition. Species whose predators merely look for deviations from the background patterns are more likely to be simply eucryptic. (6) A given defense mechanism will probably not work as well if it serves other functions. For example, the color patterns of guppies are not as cryptic as they would be if they were not also subject to sexual selection. In many cases it is possible to modify the signal so that it is less conspicuous (contractile chromatophores and folding fins), and it can be made very conspicuous during courtship. (7) Prey species with two equally dangerous predators, which differ in color discrimination and visual acuity, will have crypsis values which are compromises, and not particularly good for defense against either predator. This will be especially true if the predators hunt at different times or under different conditions. The more similar the two predators are, the better the crypsis will be. (8) If the most dangerous predator uses a particular sensory mode, then courtship signals may evolve more rapidly in other sensory modes ("private channels"). However, the predator may then switch (in evolutionary time) to the other modes. The resulting "arms race" will favor multiple sensory modes for courtship, predator defense, and predator search. Within a sensory mode, another form of "arms race" may lead to the elaboration of certain defenses, inhibiting the use of the same structures (or colors) in courtship. Much needs to be done to test these predictions.

Four speculations are also possible, and some may be testable. (1) The "arms races" might be one explanation for the origin of, as well as the use of, multiple sensory modes in all animals; predation is probably as old as the earliest organisms, and "private channels" must always have been useful. (2) Long term changes in predator species composition within a prey distribution will favor multiple defense mechanisms. (3) There may be a relationship between the efficiency of predation and the kinds of defense mechanisms used. For example, as attack efficiency increases, perhaps mechanical and chemical defenses would become more common. (4) Parasites may affect the efficiency of prey defenses with respect to predators which are the primary hosts of the parasite. Many parasites go through many hosts during their life cycles, and the transfer between hosts often depends upon predation of one host by another. The parasites affect the behavior or appearance of arthropod and mollusk hosts to make it more likely that they will be eaten, and the modifications serve to weaken the hosts's anti-predator defenses. Do parasites of fishes, amphibians and reptiles also affect their hosts' defenses?

The evolution of distastefulness, and other defenses which operate during the later predation stages, is a difficult question. For example, what favors distastefulness and warning coloration, instead of crypsis and non-distastefulness? One of the common explanations is that the aggregation of kin favors aposematism through kin selection (Harvey et al., 1982). But, aposematic and warning coloration in amphibians is not associated with gregariousness (except in some tadpoles such as *Hyla geographica*), and bright colors are often found in non-distasteful fishes. It is curious that distastefulness is rare in fishes, reptiles, and caecilian amphibians, but common in other amphibians. The reasons for this are obscure, but they might be related to microhabitat. It is most common in the more terrestrial species, and may have originally been associated with skin secretions involved in water retention. This would repay further study.

It is clear that there is a tremendous amount to be done in understanding anti-predator defense mechanisms. It is imperative that we do more comparative work on the predation-defense sequence, and study more mechanisms within the same species. There is almost nothing known about any defenses except for a few well-studied topics such as visual mimicry, and there is even less known about the integration of different defenses or the effects of joint predation by very different predators. Yet it is clearly possible to make and test predictions about these phenomena.

Acknowledgments

I thank George Lauder and two anonymous reviewers for suggestions on an earlier draft of this paper, and the National Science Foundation for many years of support of my research into anti-predator defenses.

References

Bauwens, D., and C. Thoen. 1981. Escape tactics and vulnerability to predation associated with reproduction in the lizard *Lacerta vivipara. J. Anim. Ecol.* 50: 737-743.

Bell, M. A., and T. R. Haglund. 1978. Selective predation of threespine sticklebacks (*Gasterosteus aculeatus*) by garter snakes. *Evolution* 32: 304-319.

Brodie, E. D., Jr. 1977. Salamander anti-predator postures. *Copeia* 1977: 523-535.

Brodie, E. D., Jr., and R. R. Howard. 1973. Experimental study of Batesian mimicry in the salamanders *Plethodon jordani* and *Desmognathus ocropheus. Am. Midl. Nat.* 90: 38-46.

Brodie, E. D., Jr., R. A. Nussbaum, and M. D. Giovanni. 1984. Anti-predator adaptations of Asian salamanders (Salamandridae). *Herpetologica* 40: 56-68.

Cott, H. B. 1940. *Adaptive colouration in animals.* London: Methuen.

Curio, E. 1976. *The ethology of predation.* Berlin: Springer-Verlag.

Dial, B. E., and L. C. Fitzpatrick. 1983. Lizard tail autotomy: function and energetics of post autotomy tail movement in *Scinecella lateralis. Science* 219: 391-393.

Edmunds, M. 1974. *Defense in animals: a survey of anti-predator defenses.* London: Longman.

Endler, J. A. 1978. A predator's view of animal color patterns. *Evol. Biol.* 11: 319-364.

Endler, J. A. 1981. An overview of the relationships between mimicry and crypsis. *Biol. J. Linn. Soc. Lond.* 16: 25-31.

Endler, J. A. 1983. Natural and sexual selection on color patterns in poeciliid fishes. *Env. Biol. Fish.* 9: 173-190.

Endler, J. A. 1984. Progressive background matching in moths and a quantitative measure of crypsis. *Biol. J. Linn. Soc. Lond.* 22: 187-231.

Endler, J. 1985. *Natural selection in the wild*. Princeton: Princeton Univ. Press.

Ford, E. B. 1975. *Ecological genetics*, 4th ed. London: Chapman and Hall.

Gans, C. 1974. *Biomechanics, an approach to vertebrate biology*. Philadelphia: Lippincott.

Gould, S. J., and E. S. Vrba. 1982. Exaptation--a missing term in the science of form. *Paleobiology* 8: 4-15.

Greene, H. W., and R. W. McDiarmid. 1981. Coral snake mimicry: does it occur? *Science* 213: 1207-1212.

Harosi, F. I. 1976. Spectral relations of cone pigments in goldfish. *J. Gen. Physiol.* 68: 65-80.

Harvey, P. H., J. J. Bull, M. Pemberton, and R. J. Paxton. 1982. The evolution of aposematic coloration in distasteful prey: a family model. *Am. Nat.* 119: 710-719.

Harvey, P. H., and P. J. Greenwood. 1978. Anti-predator defense strategies: some evolutionary problems. In *Behavioral ecology, an evolutionary approach*, ed. J. R. Krebs and N. B. Davies, pp. 129-151. Oxford: Blackwell Scientific Publications.

Heatwole, H. 1968. Relationship of escape behavior and camouflage in anoline lizards. *Copeia* 1968: 109-113.

Hensel, J. L., Jr., and E. D. Brodie, Jr. 1976. An experimental study of aposematic coloration in the salamander *Plethodon jordani*. *Copeia* 1976: 59-65.

Howard, R. R., and E. D. Brodie, Jr. 1973. A Batesian mimetic complex in salamanders: response of avian predators. *Herpetologica* 29: 33-41.

Huheey, J. E., and R. A. Brandon. 1974. Studies in warning coloration and mimicry. VI. Comments on the warning coloration of red efts and their presumed mimicry by red salamanders. *Herpetologica* 30: 149-155.

Johnson, J. A., and E. D. Brodie, Jr. 1975. The selective advantage of the defensive posture of the newt *Taricha torosa*. *Am. Midl. Nat.* 93: 139-148.

Levine, J. S., and E. F. MacNichol, Jr. 1979. Visual pigments in teleost fishes: effects of habitat, microhabitat, and behavior on visual system evolution. *Sensory Processes* 3: 95-131.

Lythgoe, J. N. 1979. *The ecology of vision*. Oxford: Oxford University Press.

Nowak, R. T., and E. D. Brodie, Jr. 1978. Rib penetration and associated anti-predator adaptations in the salamander *Pleurodeles waltl* (Salamandridae). *Copeia* 1978: 424-449.

Pietriwicz, A. T., and A. C. Kamil. 1981. Search images and the detection of cryptic prey: an operant approach. In *Foraging behavior: ecological, ethological, and psychological approaches*, ed. A. C. Kamil and T. D. Sargent, pp. 311-331. New York: Garland STPM Press.

Pough, H. 1974. Comments on the presumed mimicry of red efts (*Notophthalmus*) by red salamanders (*Pseudotriton*). *Herpetologica* 30: 24-27.

Silberglied, R. E., A. Aiello, and D. M. Windsor. 1980. Disruptive coloration in butterflies: lack of support in *Anartia fatima*. *Science* 209: 617-619.

Simpson, G. G. 1944. *Tempo and mode in evolution*. New York: Columbia Univ. Press.

Soan, I. D., and B. Clarke. 1973. Evidence for apostatic selection by predators using olfactory cues. *Nature* 241: 62-64.

Tilley, S. G., B. L. Lundrigan, and L. P. Brower. 1982. Erythrism and mimicry in the salamander *Plethodon cinereus*. *Herpetologica* 38: 409-417.

Vitt, L. J. 1983. Tail loss in lizards: the significance of foraging and predator escape modes. *Herpetologica* 39: 151-162.

9 Behavioral Responses of Prey Fishes During Predator-Prey Interactions

Gene S. Helfman

1. Introduction

As evidenced by the variety of disciplines represented in this volume, a diversity of successful approaches can be taken in exploring the evolution of predator-prey interactions. These approaches can be generally categorized as either indirect or direct. Indirect studies, exemplified by Chapters 2, 4, 5, and 6 (in part), often involve live or post-mortem observation and analysis of food habits or anatomical structures associated with prey capture or predator avoidance, or experimental manipulations of animals and subsequent measurement of individual growth and population responses to manipulation. The investigator then deduces the evolutionary influences that predators have had on prey and vice versa and the probable outcome of specific or hypothetical predator-prey interactions.

Direct methods, such as those discussed in Chapters 3, 6 (in part), 7, and 10, often entail ethological observations in the field or laboratory of behavioral interactions between predators and prey. Such studies have a strong historical tradition in animal behavior and have led to detailed descriptions of behavioral interactions, greatly increasing our understanding of predator-prey relationships (e.g., Curio, 1976; Edmunds, 1974; Keenleyside, 1979). More recently, behavioral phenomena associated with predator-prey interactions have become the focus of ecological studies. This behavioral-ecological approach has at its foundation evolutionary principles. Behavioral ecologists attempt to understand behavior in terms of fitness maximization, optimization, cost/benefit calculations, life history strategies, and other parameters of natural selection (e.g., Kerfoot and Sih, 1986; Morse, 1980; Taylor, 1984). This behavioral-ecological approach often involves traditional descriptive methods, in addition to extensive quantification followed by experimental manipulation. The emphasis in this chapter will be on the behavioral-ecological approach to the study of predator-prey interactions among fishes, particularly the evolved responses of prey fishes when confronted by piscivores.

Part of the fascination of predation to behavioral ecologists lies in its obvious influence on evolutionary processes. A single predator-prey interaction can have considerable influence on the fitness of a prey individual -- one mistake and a prey animal may be eliminated (Chapter 5; Helfman, 1984; Hobson et al., 1981). Such strong selection pressures should produce rather obvious adaptations. My focus will be on the potential importance of predation from the evolutionary perspective of the prey, in terms of the behavioral options open to prey fishes in response to predators (see also Chapter 8). Four themes will be emphasized, with examples from selected studies:

(1) Predation is a strong and direct determinant of much of the behavior of prey fishes. The reviewed studies show direct effects on predator avoidance behavior, reproductive behavior, and twilight activities.

(2) Prey responses are not stereotyped. Prey have the ability to adjust their anti-predator behavior to fit the particular circumstances of the interaction.

(3) Predation is non-egalitarian in nature; predation pressure falls heaviest on young animals. This fact has strong practical and theoretical implications for studying predator-prey relationships.

(4) Although predation may be an important determinant of prey fish behavior, demonstrating cause and effect will often require rigorous testing.

As I hope will be evident, the behavioral-ecological approach has contributed greatly to our understanding of predation as a biological process. However, and in analogy to a point emphasized in Chapters 2, 3, 6, and 10, the best behavioral-ecological studies of predation in fishes and other groups have been and will continue to be those that build not only on traditional ethology, but also that incorporate and integrate information obtained by functional morphologists, physiological ecologists, neurobiologists, population ecologists and evolutionary geneticists working on a variety of taxa.

2. Avoidance Behavior

2.1 Distribution

Do prey fishes shift their locale in response to predators? An individual prey animal should be able to reduce its susceptibility to predators by adjusting the place where it engages in various important activities, such as breeding and feeding. A number of studies have found that prey distributions are inversely correlated with predator distributions. For example, McKaye (1984) looked at the lekking (or male display) area of a Lake Malawi cichlid, *Cyrtocara eucinostomus*. He found that the lekking grounds were relatively shallow, probably because an abundant, predatory bagrid catfish occupied deeper water. The cichlids showed some ability to shift their distribution as a function of predation. When flocks of cormorants began feeding in the region of the lekking ground, most of the cichlids moved to deeper water and then returned within 15 minutes of the departure of the cormorants.

Such distributional discontinuities become more striking when it appears that potential prey are giving up something obvious by occurring where predators are not. Power (1983) has recently documented that loricariid catfishes in Panamanian streams occur deeper in the water than would be predicted by the distribution of their primary, and apparently limiting, algal food resources. She concluded that the distribution of the catfishes was influenced by the activities of wading birds which prey only in very shallow water. Interestingly, size differences in distribution of the prey ran counter to a general pattern for fishes, in that susceptibility to bird predation was directly related to fish size. Smaller fish were less susceptible to bird predators such as herons and kingfishers because small fish were better able to seek refuge amongst cobbles and also were smaller targets for the directed pecking attacks of the birds. Power also concluded that predatory fish in her streams preyed more heavily on small than on larger catfishes, which is the usual case for predator-prey relationships among fishes.

Although strongly suggestive, such correlation studies of distribution cannot account entirely for alternative explanations. More convincing evidence for distributional shifts caused by predation pressure comes from the experimental pond studies of Earl Werner and colleagues (Werner et al., 1983). Werner et al. placed juvenile bluegill sunfish (*Lepomis macrochirus*) of three size classes in experimental ponds that contained or lacked largemouth bass (*Micropterus salmoides*) as predators. All prey were under 75 mm. When predators were absent, all three size classes of prey occurred in the open water sections of the ponds, where more profitable food occurred. When predators were added, the smallest prey fish restricted their foraging to the vegetated regions of the pond, which had less food (Fig. 1). Foraging in this region was 1/3 what it was in open water and growth rates were reduced by 27%. When predators were present, the vulnerable size class was apparently able to shift to the less profitable but safer habitat. Larger, less vulnerable bluegill did not show similar diet and habitat shifts. These results suggest plasticity in the behavior of the prey and differential responses according to size. However, we do not know for certain if the bluegill that foraged in the vegetation when the predators

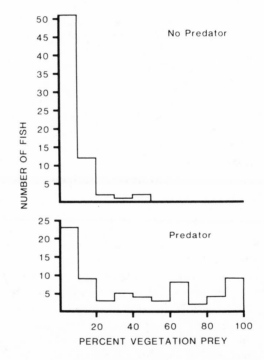

Fig. 1. Prey fish can shift their distribution and foraging patterns in response to changing predation. The number of small, vulnerable bluegill sunfish that foraged in the less profitable, vegetated region of a pond increased when predatory largemouth bass were present, as indicated by the percent of plant-associated prey in their stomachs. Vegetation generally provides a refuge from predators. In the absence of the predator, small bluegill foraged more in open water, where more profitable prey were abundant. From Werner et al. (1983).

were introduced made an active habitat switch or were individuals that had a predisposition for foraging in vegetation so that their open water brethren were eaten.

2.2 Schooling

Does schooling behavior contribute to predator avoidance and respond to variation in predation pressure? Savino and Stein (1982) looked at the relationship of schooling behavior to predation when prey fish have the opportunity to use differing densities of cover in the form of artificial plant stems. They used juvenile bluegill sunfish as prey and largemouth bass as predators. The bluegill were between 35 and 44 mm long whereas the bass were 33-37 cm. They found that:

(1) Bluegill tended to occur throughout the pool when predators were absent, but moved toward the bottom edges of the pool when the predator was present. Prey avoided the center of the pool except when plant stem densities were high, suggesting an ability to capitalize on the cover afforded by the artificial plants.

(2) At low plant stem densities, schooled bluegill were less likely to be attacked, in part because the entire school remained further from the predator than did dispersed bluegill. By being in schools bluegill were more able to keep a sufficient distance between themselves and the predator, implying increased detection of the predator.

(3) At high stem densities, bluegill behaved contrary to expectation, at least according to conventional predator-prey wisdom, which states that schooling provides protection from predators (Fig. 2). With predators present, bluegill in dense vegetation tended to disband and become immobile. At lower stem densities, they tended to school actively. At high stem densities, more fish schooled when the predator was absent than when the predator was present. This suggests that cover seeking took precedence over schooling when cover was available and predators were present. It appears that bluegill are capable of making decisions and behaviorally reduce their susceptibility to a predator by either dispersing among plants at high plant densities or by schooling at low plant densities.

Several other interesting studies on fish schooling as an antipredator response could be cited. One worth brief mention is Wolf's (1985) finding that not all species in heterospecific schools of juvenile parrotfishes (Scaridae) remain in a school when attacked by predator models. One species continued to school while another left the school and sought shelter. Apparently the anti-predator benefits that many of us are quickly willing to attribute to schooling may vary among schooling fishes.

2.3 Escape Activities

The general category of escape activities can include immobility as well as active flight and refuge seeking. Prey fishes that often employ immobility to deter predators include the sticklebacks (Gasterosteidae). Reist (1983) tested the hypothesis that brook sticklebacks (*Culaea inconstans*) with reduced anatomical defenses such as spines should exhibit better developed behavioral defenses to predators. He predicted that spineless morphs should spend more time engaged in defensive behaviors such as increased use of vegetation or earlier initiation of

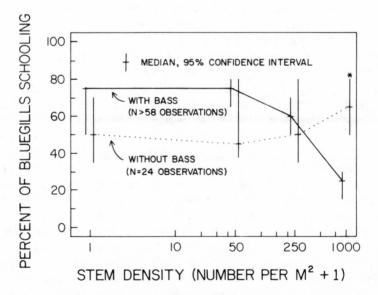

Fig. 2. Prey fish can vary their tendency to school as a function of the availability of shelter and the presence of predators. When largemouth bass were present, the percentage of bluegill swimming in schools decreased at the highest plant stem densities (solid line). No such change was observed when predators were absent (dotted line). At high plant densities, plants apparently afford better protection against predators than is achieved through school membership. From Savino and Stein (1982).

escape behaviors. He looked at three size classes of sticklebacks in the presence or absence of northern pike (*Esox lucius*) as predators.

Reist found intermorph differences in the frequency of adaptive and non-adaptive responses when predators were present or absent. The spineless morphs showed greater means and reduced variances in avoidance behavior such as cover seeking, immobility and retreating from a predator. In general, spineless morphs showed the least variation in adaptive responses, implying that "sloppiness" in anti-predator behavior is more permissible in fish with morphological defenses. Small individuals spent the least time immobile, the most time engaged in non-adaptive behaviors such as approaching a predator or continuing activity in its presence, and suffered the greatest mortality. Such ontogenetic differences in responses, with small individuals exhibiting more variable and less adaptive behavior, could have two explanations. Predators could select out behaviorally variable individuals or behaviorally inappropriate individuals among the smaller fish by preying on these incompetent animals. Alternatively, experience with predators could lead to more appropriate behavior as fish grow older -- sticklebacks may be able to learn what not to do when a pike is around. Thus competent individuals increasingly characterize the population as the fish grow older.

With respect to active predator avoidance, a popular approach to behavioral ecological questions involves attempts to calculate the relative costs and benefits of particular behavioral patterns (e.g., Chapter 5). There are obvious energetic and temporal costs associated with fleeing from a predator. According to optimization theory, prey should be able to minimize these costs and modify their behavior whenever costs can be reduced. This cost/benefit approach was applied by Coates (1980) in an experimental study of the humbug damselfish, *Dascyllus aruanus*. He postulated that time spent avoiding predators represented time lost to other important activities, like feeding or breeding or interacting socially with conspecifics. Therefore, prey fishes should be able to discriminate between predatory and non-predatory species and not waste time avoiding non-predators. Coates compared the escape responses of humbug damselfish to a series of six live predators and six live non-predatory species placed in plastic bags near the shelter site of the damselfish (Fig. 3). He found that the non-predators were basically ignored, whereas predators elicited an escape response. When predators were placed close to the coral head, the response of the prey was stronger than when predators were placed further away, whereas the response to the non-predators remained constant.

Humbug damselfish can apparently differentiate predators from non-predators and adjust their behavior accordingly, even when not under direct attack. An obvious extension of this study would be to compare the prey response to a size gradient of predators of a single species. This would test whether prey are willing to take more chances when predators are just below the threshold of effectiveness. Given the apparent size dependence of predation between fishes, can prey discriminate not only species but also recognize the relative threat posed by different sizes within a predatory species? Do prey make cost-benefit calculations based on the relative threat of a predator?

3. Reproduction

Prey fishes find themselves in something of a dilemma when reproducing, as they must spawn successfully without being eaten and, for those that engage in parental care, have to deter potential predators of their extremely vulnerable young.

3.1 Spawning Stupor

An aspect of reproductive behavior of interest with respect to predation concerns the evasive actions that spawning fishes will or will not take when under predatory attack. Many fishes spawn in predictable places at predictable times and in large numbers. It is not surprising that predators find these spawning aggregations and attack the spawning individuals. One would expect that large predators could learn the locales and times of such aggregations and exploit them. One would also expect and even predict that prey fish in spawning aggregations would be distracted and therefore especially vulnerable to predators. There is a growing literature on this topic that includes documentation of an interesting direct effect.

Not only do actively spawning fish often appear preoccupied with promoting their fitness during spawning, but fish in such spawning aggregations that are not overtly engaged in reproductive activities may also appear relatively oblivious to predators. They do not take evasive actions when predators approach them and they even approach large, threatening objects that they normally avoid, like divers with spearguns. This sort of behavior has now been documented in a number of families, including lutjanids, serranids, atherinids, mugilids, carangids, catostomids and cyprinids (e.g., Helfrich and Allen, 1975; Middaugh et al., 1981). Literature accounts on a number of other families suggest it is fairly widespread. Johannes (1981) recorded this pattern in several Palauan reef species and coined the term "spawning stupor" to describe this behavior.

Unfortunately, our information on the activities of predators and actual predator-prey interactions in spawning aggregations is largely anecdotal, because most researchers who have been fortunate enough to study spawning aggregations in the wild have been primarily interested in reproductive behavior. A systematic study of spawning stupor and its environmental and phylogenetic correlates would be fascinating. Are certain types of spawning aggregations, locales and times more likely to bring predators and spawning prey together? Is

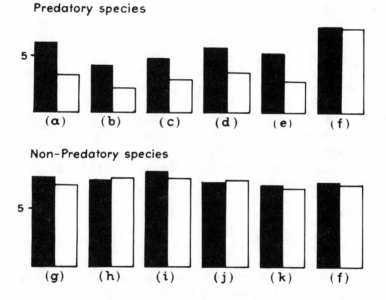

Fig. 3. Prey fishes can discriminate between predators and non-predators and adjust their avoidance behavior accordingly. The number of humbug damselfish leaving the refuge of a coral head is shown as a function of whether five predatory (a-e) or five non-predatory (g-k) fish species were placed in plastic bags and presented to the damselfish. Bar (f) represents an empty plastic bag. Damselfish responded more strongly to predators than to non-predators, and more strongly to predators 35 cm away (non-shaded bars) than to predators 70 cm away (shaded bars). No differences in response were seen between near and far non-predators, or between non-predators and an empty bag. From Coates (1980).

there a sexual asymmetry in the spawning stupor response? What are the short and long-term costs and benefits of spawning stupor to the fishes that exhibit it? Using fitness maximization theory as a guiding principle, one could predict that highly seasonal or semelparous (one time) spawners will be more likely to exhibit spawning stupor than iteroparous species that spawn repeatedly during the year or during their lives.

3.2 Parental Care

In general, parental care has been viewed as a response to high predation rates on eggs and young, although other factors such as "paternity assurance" are acknowledged as influential. Support for the idea that young fish are more subject to predation than older fish comes from the common semi-log plot used in most survivorship curves in life tables. Mortality is exceptionally high initially and then tapers off. Predators are often implicated for this loss of young fish, although real data are often lacking. A dramatic decrease in survival of young fish however has been found repeatedly in studies where parents are removed: the young are rapidly consumed by predators. But costs are associated with parental guarding; energy that could go into other activities must be spent chasing away predators. The question therefore arises as to whether parental guarding fishes can adjust their behavior to reflect changes in the threat of predators to their young?

Parental guarding species appear to be able to modify their reactions to potential and real predators as a function of the presence and vulnerability of their young. McKaye (1984) has looked at this phenomenon in cichlid fishes in both Central America and Africa and found a pattern of changing behavior in parentally guarding species. His findings from one Nicaraguan species, *Cichlasoma rostratum*, suggest that breeding fish changed the frequency and target of their agonistic attacks as a function of their nesting status:

(1) When establishing territories, aggressive attacks were directed primarily at conspecific competitors for nesting sites.

(2) After eggs were deposited in the nest, the number of attacks increased and were directed at potential egg predators such as *C. nigrofasciatum* and *Neetroplus nematopus*.

(3) As the young hatched and grew larger, aggression against small *C. nigrofasciatum*, which eat eggs but are unable to eat large fry, decreased. But aggression increased against *Gobiomorus dormitor*, an eleotrid which is primarily a predator on large fry and juveniles.

Although *Gobiomorus* constituted < 5% of the fish species present, it received > 50% of the attacks by the parental guarding fish. Also, other cichlids were generally not attacked until they came within 0.5 m of a brood, whereas *Gobiomorus* was attacked at distances > 1.0 m.

These results show the flexibility of the parental guarding prey species in response to predators as a function of the stage of the breeding cycle. The connection between changes in behavior and predatory threat are, however, correlative and some nagging doubts will remain as to real causes. The results do not control for encounter rates between parents and potential predators. Relative frequencies of parental behaviors may merely reflect the parent's encounter rate with the "appropriate" predator at that stage in the nesting cycle: the fry/juvenile predator, *Gobiomorus*, may show little interest in eggs.

4. Twilight Changeover

Another highly predictable period of vulnerability for fishes in a variety of habitats is twilight, both dawn and dusk. The work of Hobson (1968, 1972, 1975), McFarland and Munz (1976) and Helfman (1978, 1981, 1986) has clarified the behavioral, ecological and physiological factors underlying this vulnerability. Briefly, most fishes are primarily active either during the daytime or nighttime. Constraints associated with the functioning of the vertebrate eye apparently determine much of this difference: an evolutionary trade-off exists with respect to the light capture abilities of the eye, such that a retina cannot function equally well during photopic (bright light) and scotopic (dim light) conditions. Most fishes change over between activity and inactivity during both twilight periods. Conditions of illumination during twilight are also changing rapidly, such that during the period between about 15 and 30 minutes after sunset, neither the diurnal eye nor the nocturnal eye is functioning effectively.

Twilight is therefore a period of increased vulnerability of prey fishes because they are visually impaired. Predators in a number of habitat types are most active and successful at these times. They have intermediate eyes with retinal morphologies that work better under mesopic (twilight) conditions than do the eyes of diurnal and nocturnal fishes. A relative advantage is all that is required in any evolutionary situation. Predators also have a behavioral advantage in that they often lurk near the darkened bottom and strike up into the water column at prey that are backlit against the comparatively bright evening sky. Prey counter these maneuvers by getting out of the water column and ceasing activity. One apparent ontogenetic difference that may reflect relative vulnerability of different sized prey is that smaller individuals of diurnal species cease activity and seek shelter earlier in the evening and then initiate activity and leave shelter later in the morning than do their larger counterparts (Helfman, 1978, 1981; Hobson, 1972).

Another common defensive ploy of both diurnal and nocturnal prey species at twilight is to form groups to move between feeding and resting sites. Schooling has a number of anti-predator advantages that presumably apply during twilight. Several researchers have been looking into the twilight migrations of groups of juvenile grunts (Haemulidae) in St. Croix, U.S. Virgin Islands.

Grunts are nocturnally feeding fishes. They feed as solitary individuals at night on invertebrates that are usually associated with sandy or grassbed areas (Robblee and Zieman, 1984). During the daytime grunts generally occur in large daytime resting schools over coral heads or other structures. Shortly after sunset, schools migrate from the coral heads out into the grassbeds along highly predictable paths at very predictable times (Helfman and Schultz, 1984; McFarland et al., 1979; Ogden and Ehrlich, 1977). Migrations consist of fairly narrow files of fishes from which individuals split off to forage alone. The following morning, at about the same time relative to sunrise as the migration occurred relative to sunset the previous evening, the same fishes form the same schools and migrate along the same routes back to the same coral heads. These migrations are exceptionally predictable and stereotyped. From one evening to the next, when atmospheric conditions are similar, grunts will migrate within 30 seconds to one minute of when they had migrated the previous evening. The migratory route also changes by no more than a few centimeters over the initial portion of the route from one night to the next (Helfman and Wainwright, in prep.). The migrations are so stereotyped and predictable that predators such as sanddiver lizardfish, *Synodus intermedius*, frequently line up on the route near the coral head just before the grunts migrate into the grassbeds (Helfman et al., 1982; McFarland et al., 1979).

Fig. 4 (facing page). Testing for variability in the timing of twilight migrations of grunts as a function of predator activity. Resin-coated lizardfish (predator) models are attached to a monofilament line (dashed line) and placed near the migration route. When the migrating grunts approach point "X", the diver on the right jerks the monofilament line, causing the model to "attack" the lead grunt. Observer at left records data.

Peter Wainwright and I recently initiated a test of the hypothesis that the timing of migration represents a trade-off between hunger levels and predator avoidance. Grunts migrate into the grassbeds to feed. The major predator, sanddiver lizardfish, is primarily a diurnal predator which apparently ceases activity at sunset. If grunts migrated earlier they would move when the predator was most active and before their nocturnal prey had emerged from the sand. If grunts migrated later they could avoid lizardfish but would miss out on the initial emergence of their prey from the sand, some of which occupy the water column later at night. Therefore, if grunts are capable of assessing the immediate environmental situation and adjusting to it, they should migrate earlier when hungry or when predators are lacking and should migrate later if not hungry or if predators are abundant and/or active later. In our preliminary work we decided to focus on the most feasible experimental manipulation of predator activity, namely increasing the number of predators that were active and prolonging the period of predator activity on a series of evenings.

To do this we made models of the predatory lizardfish. These models were made from formalin-preserved lizardfish that had been coated with fiberglass resin (Helfman, 1983). We then tied a 2 m piece of monofilament fishing line to the lower jaw of the model and positioned several such models near the migratory route just prior to migration (Fig. 4). As the grunts migrated, one observer jerked the line such that it attacked the lead fish in a manner very similar to natural lizardfish attacks. Each time the fishes initiated migration the lead fish was attacked. The usual behavior of the grunts upon attack was to dive down and then retreat back to the coral head. Eventually the grunts migrated even though we attacked them relentlessly.

The results of these experimental predatory attacks can be seen in Figure 5. Each point represents the light level at which migration occurred; lower light levels correspond to later times. The top line suggests that the mere presence of the predatory lizardfish models on the migratory route led to a delay in initial time of migration: grunts appeared to delay migration because predators were nearby. It is fairly obvious that the eventual time of migration was delayed as a result of attacks (middle line). It also appears that on the two evenings after attacks ceased, migrations were returning to the pre-attack, earlier times and higher light levels.

We conclude that grunts are at least capable of delaying migration in the face of extended predator activity. The experiments obviously require repetition and the use of adequate controls, such as placing models of non-predatory fishes along the migratory route. We would also like to test other aspects of the idea of a trade-off between hunger and predatory activity. We could remove all predators from a reef and see whether grunts migrated earlier. We could also manipulate food availability by caging grunts for a few days and feeding them continuously or starving them to see if well-fed grunts migrated later or if starved grunts migrated earlier. But we are encouraged that what has been viewed by many researchers including ourselves as a highly stereotyped behavior is apparently open to modification and adjustment as the result of changing predation pressure.

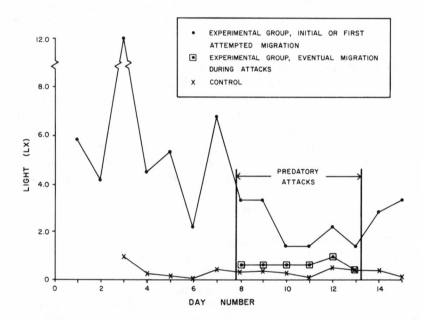

Fig. 5. Fish can vary stereotyped activities such as twilight migrations in response to changing predation. Preliminary results are shown for lizardfish launchings (Fig. 4), which tested for the effects of simulated predatory attacks on the timing of grunt migrations. Light levels at the time of migration are depicted for an experimental (attacked) and control (not attacked) school of grunts. "Controls", used to detect timing changes due to environmental factors, were larger fish that migrated later from the same coral head as the experimental fish. "Eventual migration" values are the light levels at which grunts migrated despite predatory attacks. Initial migration occurred later when predator models were lined up near the migration route prior to attack (points between the vertical lines); eventual migration was much later on the evenings of attack.

5. Caveats

Predation's star is on the rise, particularly in behavioral and ecological circles. The fact that this symposium existed with the National Science Foundation's blessing is not insignificant. One possible reason for the increased popularity of predation is that competition theory has come on hard times, for both good and bad reasons (e.g., Salt, 1984; Strong et al., 1984). The good reason is that people are reassessing much of the older work that invoked competition as an explanation for differences in morphology, behavior, distribution, feeding habits, activity times, etc. In many earlier studies, competition was assumed, uncritically, if a difference was found. Now, researchers are taking a more critical look and are concluding that certain qualifications or conditions have to be met before someone can invoke competition as an explanation.

Why this digression? Because with little difficulty, one can find many publications with the following (or a similar) statement in the abstract or discussion: "...it is hypothesized that the observed behavioral differences have arisen as a result of greater predation on the latter." I would like to offer a protocol for testing for suspected predation (Fig. 6), a protocol that might hopefully head off a violent reassessment of the importance of predation 10 years from now, a reassessment that could arise if proponents of predation make the same sorts of mistakes that many proponents of competition have made over the past 10-20 years.

The suggested protocol begins with the common situations during which someone observes a behavior and infers that it is determined by the activity of predators. The most common situation occurs when someone observes a species at two locations in a lake or reef, or in two different lakes or reefs, and notes behavioral differences that appear to correlate with differences in the presence or activity of predators. Alternatively, as Chapter 6 discusses, one may discover a natural history phenomenon, such as a distinctive behavior, that appears to be an obvious adaptation to predation or an obvious adjustment to the activity of predators.

In the next step, one documents the actual behavior and behavioral differences, assesses the activity, distribution or density of predators, and conducts statistical comparisons, such as correlation analysis, between the two to see whether or not the correlations bear out the initial surmise. Many of us have been satisfied to stop at the point of documentation and correlation and to say that it is obvious that we are looking at a predator-induced phenomenon. However, to demonstrate convincingly that predators are in fact responsible for what we have observed, it is necessary to perform some form of manipulation of the predators or the prey (see also Chapter 6).

These manipulations take two general forms. One can either remove or add predators. Predator removal can be done in a non-invasive way. Predators can be captured live and either held until the experiment is terminated or transplanted to another area. But reduction in the number of predators may or may not have the desired effect: many prey animals subjected to predation on a regular basis may not necessarily assume that because predators are not visible, predators do not exist. Many predators remain hidden from view or are rovers; prey must be on guard even when predators are not visible.

Alternatively, one can add predators. This can be done by capturing live predators and transplanting them to the study area (e.g., Power et al., 1986). This tends to be somewhat difficult, particularly in the field, because predators often have large home ranges and attempt

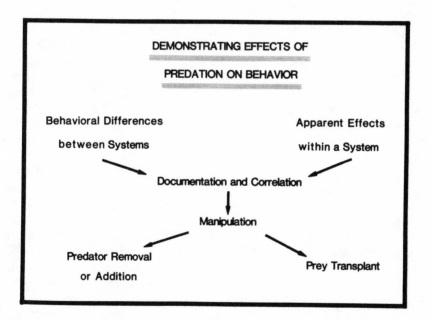

Fig. 6. Suggested protocol for testing the suspected effects of predators on fish behavior. The protocol begins with observations and documentation of behavioral differences between or within water bodies, correlated with predation differences. It progresses through experimental manipulation of predator density or reciprocal transplants of prey between areas.

to return to their home regions. Another way of adding predators is to use models, particularly if they are fairly realistic. Examples include Coates' (1980) use of live predators in plastic bags, Wolf's (1985) work on schooling, and our own work with resin-coated lizardfish. I emphasize that these models need be relatively, rather than absolutely, natural and realistic. The use of models has been widespread in studies of fish behavior, with mixed results. Most such studies are done on reproductive or territorial behavior in fishes and it is often necessary to come up with a very realistic model to elicit a fairly realistic response. It may be that fishes tend to be much more discriminating when it comes to courting or attacking models of reproductive animals than is necessary when dealing with predator-prey interactions. I suspect that most prey animals have to pay attention to anything that looks like a predator. As I mentioned earlier, one interaction between a predator and prey can have a rather striking influence on the prey individual's fitness. I think prey probably pay closer attention to our crude models of predators than they would to models of other things that are important in their lives.

The other type of manipulation that is commonly performed involves reciprocal or controlled transplants of prey between predator dense and predator scarce areas. In the field this can sometimes be difficult, although prey fishes often have smaller home ranges than predators and establish themselves more readily. We have been successful in transplanting small juvenile grunts and getting them to take up membership in new schools (Helfman and Schultz, 1984), although young animals were much easier to transplant than older conspecifics.

This outline for testing the presumed operation of predation is by no means a novel procedure. In fact, it is largely a modification of protocols currently proposed by ecologists to validate studies of presumed competition (e.g., Connell, 1984; Schoener, 1983). By using this or a similar approach, we may avoid reliving unpleasant history in ecology. We may also be able to actually say that the behavior in question is a result of predation pressures and that the behavior reflects plasticity in prey responses that result directly from the presence and/or activity of predators, without invoking the ghosts of predation past.

6. Summary and Conclusion

The studies cited here suggest three basic conclusions. First is that predation is important but that inference of the effects of predation on behavior is much more abundant than actual evidence for these effects. The work of Earl Werner and his colleagues (1983), of Savino and Stein (1982), and of Coates (1980) are good examples of studies that have adequately tested for the effects of predators on prey behavior.

The second conclusion is that predation appears to be strongest on young animals. All of the studies cited earlier either found greater vulnerability in young or small fish or used small fish to test ideas about predation. The take home message here is that to study predation one will probably have more success by concentrating on young animals.

The third conclusion is that prey fishes can adjust their behavior, both in the short term and in the long term, to the presence and activities of predators. Individuals can make immediate behavioral adjustments as a function of the particular circumstances. In the long term, prey differ in response ontogenetically, with older fish possibly being more competent, although this apparent differential in competence is more implied than demonstrated (see also Helfman, 1981; Helfman et al., 1982). Improvement in *competitive* ability with growth has been demonstrated in fishes (reviewed in Werner and Gilliam, 1984); a parallel phenomenon may apply to antipredator behavior. A remaining question here is whether an apparent increase in competence would result from (1) predators acting as selective agents and removing incompetent young prey, or (2) whether part of the flexibility of prey behavior includes the ability to learn how to avoid predators, such that increased competence resulted from individual adjustments based on prior experience. No one has to my knowledge tested for the relative contributions of selection versus learning as determinants of the apparent increased competence shown by older prey fishes. Alternatively, if predation pressure is as strong on young fishes as it appears from the literature, we might anticipate finding specific adaptations for countering predation in young fish. The large number of fishes that school only while young might be an example. We may, upon further scrutiny, fail to substantiate the observation that juvenile fishes are of limited competence with respect to antipredator adaptations. Additional documentation and manipulation are obviously needed.

Acknowledgments

I would like to thank the conveners, George Lauder and Martin Feder, for their time and efforts in producing an educational as well as enjoyable symposium. Several ideas presented here resulted from discussions with John Endler, Gary Grossman, Mary Power, and Peter Wainwright. Jim Beets, Stan Hales and the ever-present anonymous reviewer critiqued an earlier draft. Peter Wainwright and Carolyn Teragawa were indispensable during field collection of grunt data as well as refinement of the lizardfish launching methodology. Some of the field work described here was supported by the University of Georgia Research Foundation, Inc. My thanks to Jennifer Biggers for her rendition of the lizardfish launching in Figure 4, based on murky slides and verbal accounts.

References

Coates, D. 1980. The discrimination of and reactions towards predatory and non-predatory species of fish by humbug damsel fish, *Dascyllus aruanus* (Pisces, Pomacentridae). *Z. Tierpsychol.* 52: 347-354.

Connell, J. H. 1984. On the prevalence and relative importance of interspecific competition: evidence from field experiments. In *Ecology and evolutionary biology. A round table on research*, ed. G. W. Salt, pp. 81-116. Chicago: Univ. of Chicago Press.

Curio, E. 1976. *The ethology of predation*. Berlin: Springer-Verlag.

Edmunds, M. 1974. *Defense in animals*. London: Longman.

Helfman, G. S. 1978. Patterns of community structure in fishes: summary and overview. *Env. Biol. Fish.* 3: 129-148.

Helfman, G. S. 1981. Twilight activities and temporal structure in a freshwater fish community. *Can. J. Fish. Aquat. Sci.* 38: 1405-1420.

Helfman, G. S. 1983. Resin-coated fishes: a simple model technique for in situ studies of fish behavior. *Copeia* 1983: 547-549.

Helfman, G. S. 1984. School fidelity in fishes: the yellow perch pattern. *Anim. Behav.* 32: 663-672.

Helfman, G. S. 1986. Fish behavior by day, night and twilight. In *The behavior of teleost fishes*, ed. T. J. Pitcher, Ch. 14. London: Croom-Helm. In press.

Helfman, G. S., J. L. Meyer, and W. N. McFarland. 1982. The ontogeny of twilight migration patterns in grunts (Pisces: Haemulidae). *Anim. Behav.* 30: 317-326.

Helfman, G. S., and E. T. Schultz. 1984. Social transmission of behavioural traditions in a coral reef fish. *Anim. Behav.* 32: 379-384.

Helfrich, P., and P. M. Allen. 1975. Observations on the spawning of mullet, *Crenimugil crenilabis* (Forsskal) at Enewetak, Marshall Islands. *Micronesica* 11: 219-225.

Hobson, E. S. 1968. Predatory behavior of some shore fishes in the Gulf of California. *U.S. Bur. Sport Fish. Wildlife, Res. Report* 73: 1-92.

Hobson, E. S. 1972. Activity of Hawaiian reef fishes during evening and morning transitions between daylight and darkness. *U.S. Fish. Bull.* 70: 715-740.

Hobson, E. S. 1975. Feeding patterns among tropical reef fishes. *Am. Sci.* 63: 382-392.

Hobson, E. S., W. N. McFarland, and J. R. Chess. 1981. Crepuscular and nocturnal activities of Californian nearshore fishes, with consideration of their scotopic visual pigments and the photic environment. *U.S. Fish. Bull.* 79: 1-30.

Johannes, R. E. 1981. *Words of the lagoon: fishing and marine lore in the Palau District of Micronesia*. Berkeley: Univ. of Calif. Press.

Keenleyside, M. H. A. 1979. *Diversity and adaptation in fish behaviour*. Berlin: Springer-Verlag.

Kerfoot, W. C., and A. Sih, eds. 1986. *Predation: direct and indirect impacts on aquatic communities*. Hanover, N.H.: University Press of New England. In press.

McFarland, W. N., and F. W. Munz. 1976. The visible spectrum during twilight and its implications to vision. In *Light as an ecological factor: II*, ed. G. C. Evans, R. Bainbridge and O. Rackham, pp. 249-270. Oxford: Blackwell Scientific Publications.

McFarland, W. N., J. C. Ogden, and J. N. Lythgoe. 1979. The influence of light on the twilight migrations of grunts. *Env. Biol. Fish.* 4: 9-22.

McKaye, K. R. 1984. Behavioural aspects of cichlid reproductive strategies: patterns of territoriality and brood defense in Central American substratum spawners and African mouth brooders. In *Fish reproduction: strategies and tactics*, ed. G. W. Potts and R. J. Wooton, pp. 245-273. London: Academic Press.

Middaugh, D. P., G. I. Scott, and J. M. Dean. 1981. Reproductive behavior of the Atlantic silverside, *Menidia menidia* (Pisces, Atherinidae). *Env. Biol. Fish.* 7: 269-276.

Morse, D. H. 1980. *Behavioral mechanisms in ecology*. Cambridge: Harvard Univ. Press.

Ogden, J. C., and P. R. Ehrlich. 1977. The behavior of heterotypic resting schools of juvenile grunts (Pomadasyidae). *Mar. Biol.* 42: 273-280.

Power, M. E. 1983. Grazing responses of tropical freshwater fishes to different scales of variation in their food. *Env. Biol. Fish.* 9: 103-116.

Power, M. E., W. J. Matthews, and A. J. Stewart. 1986. Grazing minnows, piscivorous bass and stream algae: dynamics of a strong interaction. *Ecology,* in press.

Reist, J. D. 1983. Behavioral variation in pelvic phenotypes of brook stickleback, *Culaea inconstans,* in response to predation by northern pike, *Esox lucius. Env. Biol. Fish.* 8: 255-267.

Robblee, M. B., and J. C. Zieman. 1984. Diel variation in the fish fauna of a tropical seagrass feeding ground. *Bull. Mar. Sci.* 34: 335-345.

Salt, G. W., ed. 1984. *Ecology and evolutionary biology. A round table on research.* Chicago: Univ. of Chicago Press.

Savino, J. F., and R. A. Stein. 1982. Predator-prey interaction between largemouth bass and bluegills as influenced by simulated, submersed vegetation. *Trans. Am. Fish. Soc.* 111: 255-266.

Schoener, T. W. 1983. Field experiments on interspecific competition. *Am. Nat.* 122: 240-285.

Strong, D. R. Jr., D. Simberloff, L. G. Abele, and A. B. Thistle, eds. 1984. *Ecological communities: conceptual issues and the evidence.* Princeton: Princeton Univ. Press.

Taylor, R. J. 1984. *Predation.* New York: Chapman and Hall.

Werner, E. E., and J. F. Gilliam. 1984. The ontogenetic niche and species interactions in size-structured populations. *Ann. Rev. Ecol. Syst.* 15: 393-425.

Werner, E. E., J. F. Gilliam, D. J. Hall, and G. G. Mittelbach. 1983. An experimental test of the effect of predation risk on habitat use in fish. *Ecology* 64: 1540-1548.

Wolf, N. G. 1985. Odd fish abandon mixed species groups when threatened. *Behav. Ecol. Sociobiol.* 17: 47-52.

10 Laboratory and Field Approaches to the Study of Adaptation

Stevan J. Arnold

1. Introduction

Adaptation and selection are central concepts in the study of predator-prey relationships. During predatory pursuit or struggle the adaptations of the predator are pitted against the adaptations of prey, allowing us to visualize the biological roles of behaviors, structures and physiological processes. Hypotheses about role or adaptive significance gained from the observation of predatory struggle can give ecologically sound motivation for studies of function (Chapter 2). The role of a structure or process can also be the focus of study, for we can ask whether variation in the structure affects variation in fitness: Does the horned lizard's armature really protect against predators? In principle we could ask whether lizards with intermediate armature leave the most progeny and whether average armature in the population coincides with an optimum amount of armor. These questions are equivalent, respectively, to asking whether stabilizing selection acts on armor and whether the population is perched on an adaptive peak. We could also ask whether selection has genetic consequences. Are differences in armature inherited? Answering these questions in practice is another matter, but the answers can be considered an analysis of adaptation. Clearly the answers do not tell us everything about adaptation that we would like to know. Questions about evolutionary history, for example, are unanswered. Nevertheless we would be able to tell whether a particular population is under selection, whether it is actively evolving and whether it is in a state of local adaptation. Other approaches to the study of adaptation and selection are reviewed by Endler (1985).

The goal of this paper is to discuss the practical aspects of a strategy for analyzing adaptations. The strategy focuses on measures of whole organism performance and uses natural variation in those measures to seek correlations with behavior, physiology, morphology, on the one hand, and correlations with fitness, on the other. By performance, I mean those standardized measures of organismal ability (e.g., sprint speed, running endurance, leaping ability) that are increasingly used by physiological ecologists and functional morphologists as integrative measures of a variety of underlying physiological processes and morphological structures (e.g., Bartholomew, 1958; Bennett, 1980; Drummond, 1983; Emerson, 1978; Feder and Arnold, 1982; Greenwald, 1974; Huey and Stevenson, 1979; Webb, 1984). Throughout this paper I will use "morphology" as a short-hand for such behavioral, physiological and morphological variables.

The particular approach discussed here was developed in collaboration with A. F. Bennett, while investigating crawling performance in newborn garter snakes. Initially, we simply planned to test for differences among litters in crawling performance. As we discussed logistics, we realized that if we counted vertebral numbers on each snake, we could seek correlations with performance and so determine whether natural selection on these skeletal features (already the focus of a field study) might be mediated by locomotion. Also, we realized that if we individually marked and released the snakes after scoring performance, we might be able to measure their fitnesses through recaptures and so estimate the selective value of performance. Furthermore, because data on families were available at each step, it might be possible to analyze the inheritance of morphology, performance and fitness, as well as their hereditary correlations. The steps in this plan are outlined in Table 1, together with the types of data analysis that are possible with success at each step. We were able to proceed only as far as step five with our garter snake project because demands on our time by other field projects prevented us from mounting an intensive recapture program. Nevertheless, the plan might succeed in its entirety with the right organism, under the right circumstances.

A critical feature of the species to be studied is whether cohorts of large size can be obtained and tested, because large samples are needed to conduct the multivariate statistical analysis of data. Thus it must be possible to obtain dozens or scores of gravid females or clutches in the field or to breed the organism in captivity. In the field phase of the program, it would help to use organisms that are conspicuous and sedentary, so that recapture is feasible, and that are short-lived, so that lifetime fitness can be estimated. The virtue of the entire plan is that it integrates the laboratory study of performance with the field study of fitness. Total success could tell us, for example, whether and how strongly natural selection acts on various measures of performance. Partial success also reaps rewards. Thus if the field phase cannot be conducted, as in the garter snake case, one can nevertheless analyze the correlates of performance, as well as the inheritance of morphology and performance. The most critical step in the program is the choice of characters. In the ideal case, the character should play a key role in adaptation to a particular feature of the environment. Biomechanical analysis (e.g., Gans, 1974; Chapter 2), physiological (Chapter 6), ecological (e.g., Price et al., 1984a) or ethological studies (Curio, 1976; Chapter 6) might identify such plausibly adaptive traits. Likewise, rapid microevolution, as revealed by dramatic geographic variation within species, or a history of macroevolutionary change are also earmarks of worthy traits.

One can argue the opposite viewpoint, that selection studies should focus on traits of arbitrary adaptive value. Van Valen (1963), has suggested, for example, that traits should be chosen for study without reference to their adaptiveness so that we can gain a statistical picture of how strongly selection acts on the average character. In this view,

TABLE 1: A Scheme for Analyzing Adaptations in Natural Populations. "Morphology" Is Shorthand for Behavioral, Physiological and Structural Traits

Research Step	Data Analysis
Laboratory Phase	
1. Obtain clutches or gravid females (or captive breeding).	
2. Gestation and birth in laboratory.	
3. Evaulate performance (two or three times).	Compute repeatability of performance, genetic variances and covariances of performance measures.
4. Score morphological traits.	Analyze morphological correlates of performance (performance gradients); genetic covariances between morphology and performance.
Field Phase	
5. Release individually-marked specimens in field.	
6. Score fitness with recapture program.	Analyize correlations of morphology and performance with fitness (selection and fitness gradients); genetic variances and covariances of fitness components.
7. Re-evaluate performance and morphology at each recapture	Compute repeatability and ontogeny of morphology and performance in the field.

human fingerprint ridges, waltzing in mice and attraction of moths to light are ranked equally as candidates for selection study with *Anolis* toe lamellae, rodent habitat choice and the evasion of bats by moths. I would prefer to study selection on the latter set of traits precisely because they are likely to be adaptive and because Peterson (1983), Wecker (1963) and Roeder (1963) have given us ecological perspectives for framing hypotheses and interpreting results using these characters. Fingerprints, waltzing and attraction to lights may be adaptive, but they might be simple epiphenomena whose significance will never be illuminated by measurements of selection. Choosing traits because they are plausibly adaptive biases the sample of selection studies, leaving Van Valen's query unanswered, but the sample will produce connections between disciplines.

Computational techniques for measuring selection are reviewed by Lande and Arnold (1983), Arnold and Wade (1984 a,b), and Endler (1985). The theory underlying computations for the particular strategy discussed here is outlined by Arnold (1983).

In the following sections I turn first to the laboratory phases in the plan and then to the field phase. In each section I focus on three sets of issues: 1) variation and selection, 2) ontogeny, and 3) inheritance. These are the classical Darwinian issues in the study of adaptation. They are also the major ingredients in equations for evolutionary change in continuously distributed, polygenic traits (e.g., Lande, 1979). Consequently, quantitative genetic theory can be used to integrate these issues and to model microevolution (e.g., Arnold, 1981a; Lande, 1979; Price et al., 1984b; Schluter, 1984). In this paper, however, I focus on the practical, logistical aspects of implementing the plan outlined in Table 1, rather than on relating it to a theoretical structure.

In each section I review recent progress in studies using fish, amphibians and reptiles with a focus on traits relevant to predator-prey interactions. In most instances only a few studies have been attempted. In other cases (e.g., longitudinal studies of performance) no studies have been attempted.

2. Laboratory Phase

2.1 Variation and Selection

Direct and indirect effects of selection. Correlations among characters will induce indirect effects of selection. Thus even if selection acts directly on only one character, correlated characters will also be affected. Multivariate statistics can be used to separate the direct and indirect effects of selection (Lande and Arnold, 1983). The basic procedure is to measure the association between fitness (or performance) and one character while holding the effects of other characters statistically constant, using multiple regression analysis.

Of course, we cannot account for the effects of characters that are not included in the study. Thus if selection acts directly on an unmeasured character that is correlated with the characters in the study, we may misidentify the actual target of selection.

Garland's (1984) analysis of physiological correlates of running performance in black iguanas (*Ctenosaura similis*) illustrates the statistical technique of holding variables constant to measure direct effects on performance. Garland measured three aspects of locomotory performance (burst speed, endurance time and maximum running distance) in laboratory trials using a field collected sample of animals of varying size. Endurance, for example, was strongly correlated with size. After holding size (body mass) constant, Garland was able to account for nearly 90% of the residual variation in endurance with measures of leg muscle mass, maximum oxygen consumption, heart mass and enzyme (liver citrate synthase) activity. Individual differences in metabolism and metabolic enzyme activity accounted for 65% of the variation in maximum distance but none of the measurements accounted for variation in burst speed. Garland also used the hierarchical nature of his characters and accounted for 67% of the variation in oxygen consumption, for example, with enzymatic and organ mass variables.

Directional, stabilizing and correlational selection. Natural selection can act simultaneously on many aspects of the organism and in a variety of modes. Selection may directly affect the mean of a character (directional selection), the variance of a character (stabilizing or disruptive selection, depending on whether the variance decreases or increases), or the covariance between two characters (correlational selection). These modes of selection are not mutually exclusive alternatives. Selection might simultaneously shift the mean of one character while contracting its variance and also changing covariances with other characters.

The simultaneous and multifaceted effects of selection can be visualized as a selection surface. Selection on two characters is the easiest multiple character case to visualize. Imagine two axes, representing the two characters, and a vertical axis, representing relative fitness. A hill-shaped surface in this space would describe stabilizing selection on the two characters. The forces of directional selection will vary depending on where the population mean is located in the character space. If the mean lies precisely on the top of the hill, there is no directional selection on either character. On the other hand, if the mean lies off the hilltop, there will be directional selection on both characters, in the steepest uphill direction. Other combinations of directional, stabilizing, disruptive and correlational selection can be represented by other surfaces (pits, saddles, elliptical ridges, etc.). When the object is to describe the effects of characters on performance we can summarize the results as a performance surface in which the vertical axis is relative performance, rather than fitness.

An example of a performance surface is shown in Fig. 1. The plot shows contours of burst speed in newborn garter snakes (*Thamnophis radix*) as a function of numbers of vertebrae in the body and tail (Arnold and Bennett, 1985). The laboratory testing procedure is

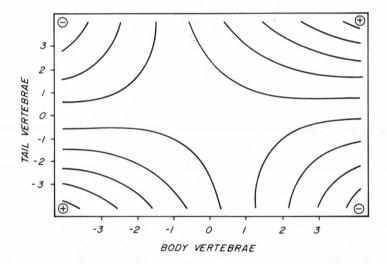

Fig. 1. Burst speed as a function of numbers of body and tail vertebrae in a sample (n = 100) of newborn garter snakes (*Thamnophis radix*) tested under uniform conditions. Burst speed is shown as contour lines and increases towards the upper right-hand corner and towards the lower left-hand corner. The performance surface was drawn using coefficients from a multiple, curvilinear regression of burst speed on numbers of body vertebrae and tail vertebrae, squared number of body vertebrae, squared number of tail vertebrae and the product of body and tail vertebral numbers. The bivariate mean of the population is located at the center of the plot. Numbers of body and tail vertebrae are on a logarithmic scale in units of phenotypic standard deviation.

described by Arnold and Bennett (1984). We were unable to detect statistically significant directional or stabilizing (disruptive) selection on either character. Our analysis did, however, reveal significant correlational selection and it is that selection coefficient that largely determines the surface portrayed in Fig. 1. In other words, the set of snakes showing the fastest burst speeds also had a higher correlation between body and tail vertebral numbers. Thus good performance was associated with the relative proportion of body and tail vertebral numbers rather than the absolute numbers in either region of the vertebral column. Swain and Lindsey (1984) staged encounters between sticklebacks (*Gasterosteus aculeatus*) and predatory sunfish (*Lepomis gibbosus*) and found directional selection for low vertebral counts. Such staged trials between predators and prey could be used to assess the stabilizing and correlational effects of selection on the attributes of both predators and prey.

2.2 Ontogeny

The repeatability and ontogeny of performance are critical issues that affect the prospects and techniques for analyzing both selection and inheritance. Repeatability is simply the consistency of individual performance. It is conveniently measured as correlation between trials, or (especially if there are more than two trials) as an intraclass correlation, estimated by analysis of variance (Falconer, 1981). In contrast to the random fluctuations in performance between trials close together in time, performance may also change progressively with age. Such individual histories can be described by ontogenetic trajectories or lines fitted to a plot of an individual's performance as a function of age, using linear or curvilinear regression (e.g., Cock, 1966).

Short-term fluctuation. Performance may vary from trial to trial without any pattern. Such fluctuations may reflect measurement error as well as intangible changes in physiology. Motivation, for example, is thought to be a leading cause of variation in performance among world-class athletes. Despite such random variation, there may be consistent differences among individuals. Thus lizards (*Sceloporus occidentalis*) maintained a consistent ranking in running performance in laboratory trials and relative performance, independent of body temperature (Bennett, 1980). Repeatability in garter snake (*Thamnophis radix*) locomotory performance, measured as the correlation between trials on successive days, ranged from $r = 0.55$ ($n = 139$) for endurance (total distance crawled) to $r = 0.71$ ($n = 39$) for burst speed (Arnold and

Bennett, 1985). Repeatability of antipredatory displays in these snakes fell in the same range (Arnold and Bennett, 1984).

Shaffer and Lauder (1985a,b) made repeated measurements of head movements and electromyographic responses during prey capture in ambystomatid salamanders. Seven aspects of head movement (e.g., cranial elevation, maximum gape) all showed significant individual differences, with an average repeatability of 0.37. Most of the eleven electromyographic variables showed significant individual variation (average repeatability = 0.31) and many showed significant differences between trials on different days. The good news from these studies is that conspecifics differ in prey catching movements. The sobering message is that movement repeatabilities may be in the low range.

Repeatability can be used as a criterion for deciding on testing protocols. If repeatability is so low that there are no significant differences among individuals, then there is no hope of detecting selection or analyzing inheritance. Improving the scoring protocol may require less effort and expense than boosting sample sizes in order to compensate for low repeatability. Another solution is to make multiple measurements on each individual and use the average or best score. Falconer (1981) discusses the gain from such multiple measurement; improvement is asymptotic with effort.

Long-term trajectories. Performance undoubtedly changes with age (and size) in many taxa and may reflect ontogenetic niche shifts, but the phenomenon has been little studied (Garland, 1985; Pough and Kamel, 1984; Werner and Gilliam, 1984). Pough (1977, 1978), for example, showed that juvenile natricine snakes had much less endurance capacity than adults. Because of such ontogenetic trends, it is crucial to control for age and size in studies of correlates of performance or in genetic analysis. One powerful approach is to make measurements of individuals of known age at regular intervals. Then individual trajectories can be used to describe. performance. In the happy event that all trajectories are parallel, we would need only one measurement of each individual, so long as we knew its age. Individuals, however, are likely to vary in the slopes as well as the elevation of their trajectories. In this case one needs to follow individual performance histories.

2.3 Inheritance

Behavioral, physiological and morphological traits, as well as performance, are likely to be influenced by numerous genes. The body of theory developed to analyze the inheritance of such polygenic traits is called quantitative genetics, which traces its ancestry to seminal papers by Weinberg (1909), Fisher (1918) and Wright (1921). In outline, the theory uses quantitative measures of resemblance among relatives in, say, performance to estimate two critical kinds of genetic parameters, additive genetic variances and covariances. These genetic parameters reflect the additive effects of genes that cause heritable resemblance between offspring and their parents. Consequently, these parameters enable us to predict traits in the next generation from selected parents in the preceding generation, and so they play a key role in equations for the evolution of polygenic traits. Falconer (1981) gives a lucid introduction to the discipline. Kempthorne (1957), Bulmer (1980), Mather and Jinks (1977), and Pollack et al. (1977) are good sources for information on special topics. Lande (1979, 1982) discusses applications to evolutionary problems.

The basic idea in applying quantitative genetics is to use replicated combinations of relatives (e.g., sets of parents and their offspring) to estimate genetic parameters. The mapping between phenotypic resemblance and the underlying genetic parameters has been worked out for most common kinds of relationships (e.g., Kempthorne, 1957).

Some kinds of relatives give better genetic information than others. If we measure performance in broods of a taxon with an unknown system of paternity, for example, we will not know whether all members of a brood are full-sibs or whether maternal half-sibs are present as well. We can test for differences among broods in performance (say, with an analysis of variance) and so make a rough test for heritable variation, but family resemblance might also be due to the environment that is common to families and to non-additive genetic effects.

Bennett's (1980) discovery of consistent individual differences in the running performance of *Sceloporus* lizards has encouraged the next stage in genetic analysis of reptiles, tests for significant variation among broods. Thus van Berkum and Tsuji (1985) have found significant family differences in burst speed in a population of *Sceloporus occidentalis*. Andrews and Pough (personal communication), likewise, found differences in locomotory performance among litters of the lizard *Chalcides ocellatus*.

Broods of full-sibs provide slightly better information. Because the relationship of littermates is known, we can place a bound on genetic variance or heritability. Heritability is simply a standardized genetic variance (the ratio of additive genetic variance to total phenotypic variance). Still, however, we can only place an upper bound on the estimate because environmental and nongenetic effects are confounded with additive gene effects in producing similarity between full-sibs.

Studies of garter snake (*Thamnophis*) populations have revealed heritable variation in the prey recognition behavior, antipredatory displays and locomotory performance of newborn animals. All studies to date have analyzed variation within and among litters of full-sibs, so the heritability estimates are upper bounds. Ayres and Arnold (1983) review the substantial evidence that littermates are full-sibs and report heritability estimates in the range 0.31-0.54 for slug-eating scores in two populations of *T. elegans*. These same populations showed heritable variation in chemoreceptive (tongue-flicking) response to odors from slugs and a variety of other natural prey. Analysis of genetic covariances in chemoreception revealed genetic couplings between responses to taxonomically related prey but there were also some startling exceptions. Responses to slugs and leeches, for example, were genetically correlated (Arnold, 1981a,b).

Maternal effects probably did not inflate the heritability estimates in these garter snake studies. Burghardt (1971) failed to influence juvenile chemoreceptive responses to prey by manipulating maternal diet during gestation, and Arnold (1981c) did not detect maternal effects in crosses between populations. Furthermore, littermates were separated and individually housed shortly after birth, disrupting common family effects. This same postnatal protocol was followed in Arnold and Bennett's (1984) study of antipredator displays in newborn *T. radix*. This study reported a heritability of 0.37 for single trials (heritability = 0.45, 95% confidence limits = 0.19-0.97 for the average of two trials). Nevertheless, the possibility of environmental contributions to family resemblance can never be completely excluded with only full-sib data. The prospects for inflated heritability estimates are great for some characters, such as garter snake birth weight (heritability estimates for *T. elegans* are greater than one) and smaller for other traits, such as chemoreception scores. Despite their deficiencies, the full-sib studies are defensible as first generation attempts that suggest that very considerable levels of genetic variation exist in natural populations. The suggestion should now be tested with breeding designs that yield better genetic information.

The best estimates of genetic parameters are achieved by breeding designs that produce two or more levels of relationship. The easiest of these designs to implement is sets of mothers and their offspring. Heritability is estimated by calculating the slope of the statistical regression of offspring phenotype on mother's phenotype. Heritability equals twice the regression slope. Equating twice the slope with heritability assumes that environmental effects do not contribute to the resemblance between parents and offspring. In addition, performance should be measured at the same age and under the same test conditions in both mothers and offspring. The most convenient test protocol, however, is to measure performance in mothers and offspring at the same time. With such data, the regression of offspring on mother's performance cannot be used to compute a heritability, unless performance is age-invariant, but they can be used to compute a genetic covariance or correlation between adult and juvenile performance.

The technique of estimating genetic variances and covariances with parent-offspring data can be illustrated with the example of vertebral number inheritance in garter snakes. Vertebral numbers in the body and tail are fixed before birth and remain constant in an individual through postnatal life (except for loss of tail tip, which is easily recognized). Because the two counts are age-invariant, the same traits can be conveniently scored in mothers and offspring at the same time, shortly after birth. The scores can be made by external counts of ventral and subcaudal scales, because of a one-to-one correspondence with body and tail vertebrae, respectively (Alexander and Gans, 1966; Voris, 1975). Heritabilities of the body and tail counts are 0.59 ± 0.19 standard error and 0.57 ± 0.18 standard error, respectively, based on regressions of about 500 offspring *T. elegans* on their 50 mothers in each of the two populations. The genetic correlation between body and tail vertebral counts can be estimated in two ways; by regressing offspring tail counts on mother's body vertebrae and by regressing offspring body counts on mother's tail vertebrae. Both estimates give the same surprising result: tail and body vertebral numbers are not genetically correlated. Even the phenotypic correlation is low ($r = 0.14$; $p < 0.02$).

Heritability can also be assessed using data on offspring and both of their parents. Greater statistical power and some gain in information is achieved by knowing the phenotypes of fathers as well as mothers. A greater slope for the offspring-mother regression compared to the offspring-father regression, for example, indicates maternal effects. Normally, however, captive breeding will be required to know paternity with confidence. If captive breeding can be done, other designs give more information than the offspring-parent regression approach. These designs also have the convenient feature that the phenotypes of parents are not measured.

The maternal full-sib/paternal half-sib design should be considered if captive breeding is feasible. The design entails breeding a small harem of females to each of several males. The rationale for the design is that difference among the offspring of the males will be due to additive gene effects if brood environments are controlled or randomized and if each male sires offspring from a comparable (large) sample of females. Under these conditions, genetic variance is equivalent to four times the among-sire component of variance (Falconer, 1981).

The decisions of how many sires to breed, the size of harems and how many littermates to score can be made using the criterion that the standard error of heritability should be minimized. Robertson (1959) found that in the absence of preliminary information about heritability and if approximately 1200 offspring can be scored, then a good general rule is to breed three or four females to each of 30-40 sires and score traits on about ten offspring in each full-sib family. An alternative plan should be pursued if nongenetic causes of resemblance among full-sibs are likely (pre- and postnatal maternal effects and common environment effects). In such cases the variation among half-sib families gives the most reliable information about heritability and genetic correlation. A good plan would be to score traits on 20-30 offspring from each of 40-60 sires, with each sire bred to multiple females (Roberston, 1959).

More complicated breeding designs should be considered if it is feasible artificially to fertilize ova, as in some amphibians and fish. The major gain from such design is that the magnitude of maternal effects can be estimated, as well as genetic parameters. One powerful design, for example, is to fertilize the ova of each female with the sperm from a series of males, while using the sperm from each male to fertilize the ova of several females. The rationale and details of such factorial designs are discussed by Cockerham (1963). Computations and interpretation are much simplified if the design is balanced and orthogonal. Special designs to elucidate maternal effects and their inheritance are detailed by Eisen (1967).

Recent work on the inheritance of flight performance in insects should encourage genetical studies of performance in vertebrates. Curtsinger and Laurie-Ahlberg (1981) characterized differences in flying performance among 21 lines of *Drosophila melanogaster* extracted from natural populations and differing only in chromosome 2. Heritability estimates ranged from 0.16 for wing stroke amplitude to 0.44 for inertial power. These are "broad sense" heritability estimates that include nonadditive genetic variance components.

3. Field Phase

3.1 Variation and Selection

Field measurement of fitness. Because selection is a statistical relationship between characters and fitness, accurate measurement of fitness is crucial. Ornithologists, following the pioneering studies of David Lack, have had the most success among vertebrate biologists in field measurement of fitness (e.g., McGregor et al., 1981; Noodwijk, 1982; Smith, 1985). Success has been achieved, for example, by banding entire breeding populations of passerines on small islands or forest tracts, constantly monitoring all nests during the breeding season and scoring laying and fledgling success. Intensive, long-term programs of this kind have yielded tallies of lifetime reproductive success for samples ranging in size from dozens to hundreds of individuals (reviewed in Clutton-Brock, 1985). The secretive habits of many fish, amphibians and reptiles frustrate measurement of fitness. Nevertheless, major components of fitness can be determined in many species, permitting measurement of particular types of selection as well as partial resolution of total selection. Thus clutch size can be determined in turtles by radiography (Gibbons and Greene, 1979), male mating success can be assessed by regularly counting clutches in the territories of male sculpins (Downhower and Brown, 1980) and frogs (e.g., Howard, 1981; Kluge, 1981) and survivorship can be established by intensive mark-recapture work (e.g., Fox, 1983; Tinkle, 1967). The current challenge is to adapt techniques developed for demographic studies, in which the focus was parameters of population growth, for use in studies of fitness and its components, in which the focus is on individual differences.

Longitudinal estimates of fitness generally provide better information than cross-sectional estimates (Lande and Arnold, 1983). In the ideal longitudinal approach, a cohort is followed from birth throughout life and the total fitness of every individual is scored. The approach is difficult because every animal must be individually recognizable and under constant surveillance, or at least monitored so closely that all progeny can be properly assigned to their parents. The advantages of the longitudinal approach are that inconsistencies of selection and the environment do not compromise the results, and selection on age-dependent traits can be measured. In the cross-sectional approach, the population is sampled at a single interval of time and the histories of individuals are not followed. Thus selection might be estimated, for example, by contrasting juvenile and adult age-classes. The results, however, are predicated on the assumption that the current juvenile class closely resembles the juvenile cohort from which present day adults were derived (e.g., comparable environmental effects on phenotypes) and only age-invariant traits can be analyzed.

Morphological correlates of fitness. Longitudinal studies of selection have hardly ever been attempted with amphibians, fish and reptiles. In one of the few studies of this kind, Fox (1975) followed a marked cohort of about 250 juvenile lizards (*Uta stansburiana*) for about 8 months and measured mortality selection on age-invariant scale counts. Fox (1975, 1983) found statistically significant stabilizing selection (variance contraction) on several scale characters but was unable to resolve selection on several other counts. Ferguson and Fox (1984) also employed the longitudinal approach with *U. stansburiana* and detected significant directional selection on hatchling size.

In contrast, several cross-sectional studies of selection on reptile scale counts have been conducted, with the general result that variance often contracts from juvenile to adult cohorts (Beatson, 1976; Dunn, 1942; Hecht, 1952; Inger, 1943; Rand, 1954; but see Klauber, 1945).

Performance and fitness in the field. The adaptive value of performance is an almost completely unexplored issue. Even though most measures of performance seem plausibly adaptive, the goal of measuring selection on performance should not be viewed as merely an attempt to validate one's intuition. Different types of performance might be in conflict, for example. Selective conflicts could be revealed by measuring correlational selection.

One expedient approach to measuring selection on performance is to combine performance measurement with an ongoing demographic study that has already succeeded in scoring the fitness of individuals in the population (R. B. Huey, personal communication). Thus McGregor et al. (1981) juxtaposed individual measures of song repertoire size with ongoing studies of lifetime fitness in great tits (*Parus major*) and so were able to detect stabilizing selection on the behavioral trait. Trevor Price (personal communication) has shown that it is feasible to measure performance in the field, as well as its morphological correlates. By carefully following and observing marked finches (*Geospiza fortis*) in the Galapagos, Price was able to score their success as they foraged on a resource base consisting of only a few seed types. Price found significant correlations between bill dimensions and seed foraging performance.

Territory size may usefully be considered a type of performance, a variable interposed between morphology and fitness. Treating territory size in this way, we can measure: 1) effects of morphology on territory size, 2) effects of territory size on fitness and 3) effects of morphology on fitness not mediated via territory size. If, on the other hand, we treat territory size as just another character (e.g., Price, 1984), we may exclude the possibility that selection acts on a character via its effects on territoriality.

3.2 Ontogeny

Performance and fitness trajectories. Two important issues in field studies are individual consistency and ontogenetic trajectory. Both issues could be approached by measuring performance, for example, at each sighting or recapture. A similar approach can be applied to fitness components, such as annual fertility or clutch size, with a repeating feature. Houck et al. (1985) illustrate measurement of repeatability of a fitness component, namely mating success, in laboratory trials.

3.3 Inheritance

Fitness genetics. The inheritance of lifetime fitness and its components in natural populations of fish, amphibians and reptiles is a virtually unexplored field. The studies conducted so far have primarily focused on tests for genetic differences between populations in growth rate and on qualitative tests for genetic variation within population rather than on parameter estimation (Berven, 1982; Reznick, 1982; Stearns, 1983; Travis, 1983). Recent progress using other taxonomic groups is reviewed by Dingle and Hegmann (1982).

A major unresolved issue is the question of whether there are tradeoffs between fitness components that would be revealed by negative genetic correlations. Genetic variance for total lifetime fitness should be nonexistent in equilibrium populations (Fisher, 1958). Nevertheless, any or all components of fitness might have non-zero genetic variances (Lande, 1982). The result is not surprising, if we remember that the variance of a sum is composed of both the variances of its parts and the covariances between the parts. If some of the genetic covariances between fitness components are negative, they could cancel the contributions from fitness component variance (which must be non-negative), and so yield zero genetic variance for total fitness.

Falconer (1981) has described a pleiotropic evolutionary process that produces negative genetic covariances between fitness components. Genes with only positive effects on either or both of two fitness components will eventually be fixed in the population. In contrast, a gene with a positive effect on one fitness component and a negative effect on the other component may be held at intermediate frequencies. Genes of this type, with antagonistic pleiotropic effects, will be the major source of genetic covariance between fitness components and that covariance will tend to be negative. The prediction of such negative covariances has been confirmed by some studies (e.g., Rose and Charlesworth, 1981) but tests are still rare. An outstanding challenge is to test the prediction of negative genetic covariances for fitness components measured in nature.

A major practical problem confronting genetical studies of fitness is that at least one full generation of surveillance is required. Unlike the scoring of age-invariant traits, such as scale counts, which can be done immediately after birth, scoring total fitness requires following individuals for a lifetime. Furthermore, because fitness and its components are often sensitive to maternal and other environmental effects, sophisticated multi-generation breeding and cross-fostering designs are highly desirable. Thus the genetical study of fitness, especially a field study, is a major undertaking and most practical with short-lived organisms. Some recent progress in laboratory studies is summarized in Dingle and Hegmann (1982).

Genetic covariance between performance and fitness components. A study that succeeds in measuring major components of fitness can provide a test for genetic coupling between performance and fitness components, if the data are family-structured. Such genetic covariances reflect pleiotropic gene action (i.e., segregating genes that affect both performance and fitness) and linkage disequilibrium (i.e., non-random associations of alleles affecting performance with alleles at other loci that affect fitness) and provide genetic evidence for a relationship between performance and fitness. Likewise, genetic covariances between morphology and fitness components are important evolutionary parameters. A study that provides estimates of total fitness might permit computation of genetic covariances between performance (or morphology) and total fitness, but we expect these covariances to be zero in a population at genetic equilibrium (Crow and Nagylaki, 1976).

4. Summary

Because behaviorists, morphologists, systematists and physiological ecologists usually study different kinds of traits, we miss many important connections between our fields. A natural way to build bridges is to include different sets of traits in one study (e.g., combine meristic and morphometric traits with whole animal measurement of performance or with behavioral reactions to prey). Do meristics and morphometrics affect locomotory performance, for example? Inheritance and selection on these same traits can be studied using the tools of quantitative genetics and ecology. Are the traits used by systematists heritable? Is performance under directional and/or stabilizing selection in nature? By combining studies of inheritance and selection, we can determine whether selection has genetic consequences. It helps to have an organism that readily delivers young or breeds in the lab, that can be individually marked and recaptured and that permits large sample sizes for both laboratory and field operations (e.g., natricine snakes, ambystomatid salamanders). Against the promise of finding new connections between fields, one must weigh the logistic difficulties of large-scale breeding and mark/recapture work and the possibility that both genetic and selection effects will be weak and, so, hard to detect.

Acknowledgments

I am grateful to A. F. Bennett, J. A. Cheverud, S. Emerson, M. E. Feder, T. Garland Jr., R. B. Huey, H. B. Shaffer, J. S. Tsuji and F. H. van Berkum for discussion and comments on the manuscript. The preparation of the manuscript was supported by N.S.F. grant BSR 81-11489 and by U.S. Public Health grant 1-K04-HD00392-01.

References

Alexander, A. A., and C. Gans. 1966. The pattern of dermal-vertebral correlation in snakes and amphisbaenians. *Zool. Mededeelingen* 41: 171-190.

Arnold, S. J. 1981a. The microevolution of feeding behavior. In *Foraging behavior: ecological, ethological and psychological approaches*, ed. A. Kamil and T. Sargent, pp. 409-453. New York: Garland STPM Press.

Arnold, S. J. 1981b. Behavioral variation in natural populations. I. Phenotypic, genetic and environmental correlations between chemoreceptive responses to prey in the garter snake, *Thamnophis elegans. Evolution* 35: 489-509.

Arnold, S. J. 1981c. Behavioral variation in natural populations. II. The inheritance of feeding response in crosses between geographic races of the garter snake, *Thamnophis elegans. Evolution* 35: 510-515.

Arnold, S. J. 1983. Morphology, performance and fitness. *Am. Zool.* 23: 347-361.

Arnold, S. J., and A. F. Bennett. 1984. Behavioural variation in natural populations. III. Antipredator displays in the garter snake *Thamnophis radix. Anim. Behav.* 32: 1108-1118.

Arnold, S. J., and A. F. Bennett. 1985. Behavioural variation in natural populations. V. Morphological correlates of locomotory performance in the garter snake *Thamnophis radix.* Manuscript.

Arnold, S. J., and M. J. Wade. 1984a. On the measurement of natural and sexual selection: theory. *Evolution* 38: 709-719.

Arnold, S. J., and M. J. Wade. 1984b. On the measurement of natural and sexual selection: applications. *Evolution* 38: 720-734.

Ayres, F. A., and S. J. Arnold. 1983. Behavioural variation in natural populations. IV. Mendelian models and heritability of a feeding response in the garter snake, *Thamnophis elegans. Heredity* 51: 405-413.

Bartholomew, G. A. 1958. The role of physiology in the distribution of terrestrial vertebrates. In *Zoogeography*, ed. C. L Hubbs, pp. 81-95. Washington, D. C.: Amer. Assoc. Adv. Sci.

Beatson, R. R. 1976. Environmental and genetical correlates of disruptive coloration in the water snake, *Natrix s. sipedon. Evolution* 30: 241-252.

Bennett, A. F. 1980. The thermal dependence of lizard behaviour. *Anim. Behav.* 28: 752-762.

Berven, K. A. 1982. The genetic basis of altitudinal variation in the wood frog *Rana sylvatica.* I. An experimental analysis of life history traits. *Evolution* 36: 962-983.

Bulmer, M. G. 1980. *The mathematical theory of quantitative genetics.* Oxford: Clarendon Press.

Burghardt, G. M. 1971. Chemical-cue preferences of newborn snakes: influence of prenatal maternal experience. *Science* 171: 921-923.

Clutton-Brock, T. H., ed. 1985. *Reproductive Success.* Chicago: Univ. of Chicago Press.

Cock, A. G. 1966. Genetical aspects of metrical growth and form in animals. *Quart. Rev. Biol.* 41: 131-190.

Cockerham, C. C. 1963. Estimation of genetic variances. In *Statistical genetics and plant breeding*, ed. W. D. Hanson and H. F. Robinson, pp. 53-94. National Academy of Science: National Research Council Publ. No. 982.

Crow, J. F., and T. Nagylaki. 1976. The rate of change of a character correlated with fitness. *Am. Nat.* 110: 207-213.

Curio, E. 1976. *The ethology of predation.* New York: Springer-Verlag.

Curtsinger, J. W., and C. C. Laurie-Ahlberg. 1981. Genetic variability of flight metabolism in *Drosophila melanogaster.* I. Characterization of power output during tethered flight. *Genetics* 98: 549-564.

Dingle, H. and J. P. Hegmann. 1982. *Evolution and genetics of life histories.* New York: Springer-Verlag.

Downhower, J. F. and L. Brown. 1980. Mate preferences of female mottled sculpins, *Cottus bairdi. Anim. Behav.* 28: 728-734.

Drummond, H. 1983. Aquatic foraging in garter snakes: a comparison of specialists and generalists. *Behaviour* 86: 1-30.

Dunn, E. R. 1942. Survival value of varietal characters in snakes. *Am. Nat.* 76: 104-109.

Eisen, E. J. 1967. Mating designs for estimating direct and maternal genetic variances and maternal covariances. *Can. J. Genet. Cytol.* 9: 13-22.

Emerson, S. 1978. Allometry and jumping in frogs: helping the twain to meet. *Evolution* 32: 551-564.

Endler, J. 1985. *Natural selection in the wild.* Princeton: Princeton Univ. Press.

Falconer, D. S. 1981. *Introduction to quantitative genetics,* 2d ed. London: Longman.

Feder, M. E., and S. J. Arnold. 1982. Anaerobic metabolism and behavior during predatory encounters between snakes (*Thamnophis elegans*) and salamanders (*Plethodon jordani*) *Oecologia* 53: 93-97.

Ferguson, E. W., and S. F. Fox. 1984. Annual variation of survival advantage of large juvenile side-blotched lizards, *Uta stansburiana*: its causes and evolutionary significance. *Evolution* 38: 342-349.

Fisher, R. A. 1918. The correlation between relatives on the supposition of Mendelian inheritance. *Trans. Roy. Soc. Edinb.* 52: 399-433.

Fisher, R. A. 1958. *The genetical theory of natural selection,* 2d ed. New York: Dover.

Fox, S. F. 1975. Natural selection on morphological phenotypes of the lizard *Uta stansburiana. Evolution* 29: 95-107.

Fox, S. F. 1983. Fitness, home-range quality, and aggression in *Uta stansburiana.* In *Lizard ecology,* ed. R. B. Huey, E. Pianka, and T. Schoener, pp. 149-168. Cambridge: Harvard Univ. Press.

Gans, C. 1974. *Biomechanics, an approach to vertebrate biology.* Philadelphia: Lippincott.

Garland, T., Jr. 1984. Physiological correlates of locomotory performance in a lizard: an allometric approach. *Am. J. Physiol.* 247: R806-R815.

Garland, T. Jr. 1985. Ontogenetic and individual variation in size, shape, and speed in the Australian agamid lizard *Amphibolurus nuchalis. J. Zool.*, in press.

Gibbons, J. W., and J. L. Greene. 1979. X-ray photography: a technique to determine reproductive patterns of freshwater turtles. *Herpetologica* 35: 86-89.

Greenwald, O. E. 1974. Thermal dependence of striking and prey capture by gopher snakes. *Copeia* 1974: 141-148.

Hecht, M. K. 1952. Natural selection in the lizard genus *Aristelliger. Evolution* 6: 112-124.

Houck, L. D., S. J. Arnold, and R. A. Thisted. 1985. A statistical study of mate choice: sexual selection in a plethodontid salamander (*Desmognathus ochrophaeus*). *Evolution* 39: 370-386.

Howard, R. D. 1981. Male age size distribution and male mating success in bullfrogs. In *Natural selection and social behavior,* ed. R. D. Alexander and D. W. Tinkle, pp. 61-77. New York: Chiron Press.

Huey, R. B., and R. D. Stevenson. 1979. Integrating thermal physiology and ecology of ectotherms: a discussion of approaches. *Am. Zool.* 19: 357-366.

Inger, R. F. 1943. Further notes on differential selection of variant juvenile snakes. *Am. Nat.* 77: 87-90.

Kempthorne, O. 1957. *An introduction to genetic statistics.* New York: John Wiley.

Klauber, L. M. 1945. Herpetological correlations. I. Correlations in homogeneous populations. *Bull. Zool. Soc. San Diego* 21: 5-101.

Kluge, A. G. 1981. The life history, social organization, and parental behavior of *Hyla rosenbergi* Boulenger, a nest-building gladiator frog. *Misc. Publ. Mus. Zool. Univ. Mich.* 160: 1-170.

Lande, R. 1979. Quantitative genetic analysis of multivariate evolution, applied to brain: body size allometry. *Evolution* 33: 402-416.

Lande, R. 1982. A quantitative genetic theory of life history evolution. *Ecology* 62: 607-615.

Lande, R., and S. J. Arnold. 1983. The measurement of selection on correlated characters. *Evolution* 37: 1210-1226.

Mather, K., and J. L. Jinks. 1977. *Introduction to biometrical genetics.* London: Chapman and Hall.

McGregor, P. K., J. R. Krebs, and C. M. Perrins. 1981. Song repertoires and lifetime reproductive success in the great tit (*Parus major*). *Am. Nat.* 118: 149-159.

Noodwijk, A. J. van. 1982. Genetic variation in life history traits in natural populations of birds. In *Evolution and the genetics of populations*, ed. S. D. Jayakar and L. Zonta, pp. 141-152. *Atti. Ass. Genet. Ital.*, Vol. 29 (Suppl).

Peterson, J. A. 1983. The evolution of the subdigital pad of *Anolis*. 2. Comparisons among iguanid genera related to the anolines and a view from outside the radiation. J. Herpetol. 17: 371-397.

Pollack, E., O. Kempthorne, and T. B. Bailey, Jr. 1977. *Proc. Internatl. Conf. on Quantitative Genetics.* Ames, Iowa: Iowa State Univ. Press.

Pough, F. H. 1977. Ontogenetic change in blood transport capacity and endurance in garter snakes (*Thamnophis sirtalis*). *J. Comp. Physiol.* 116: 337-345.

Pough, F. H. 1978. Ontogenetic changes in endurance in water snakes (*Natrix sipedon*): physiological correlates and ecological consequences. *Copeia* 1978: 69-75.

Pough, F. H., and S. Kamel. 1984. Post-metamorphic change in activity metabolism of anurans in relation to life history. *Oecologia* 65: 138-144.

Price, T. D. 1984. Sexual selection on body size, territory and plumage variables in a population of Darwin's finches. *Evolution* 38: 327-341.

Price, T. D., P. R. Grant, and P. T. Boag. 1984a. Genetic changes in the morphological differentiation of Darwin's ground finches. In *Population biology and evolution*, ed. K. Wohrmann and V. Loeschcke, pp. 49-66. Berlin: Springer-Verlag.

Price, T. D., P. R. Grant, H. L. Gibbs, and P. T. Boag. 1984b. Recurrent patterns of natural selection in a population of Darwin's finches. *Nature* 309: 787-789.

Rand, A. S. 1954. Variation and predator pressure in an island and a mainland population of lizards. *Copeia* 4: 260-262.

Reznick, D. 1982. The impact of predation on life history evolution in Trinidadian guppies: Genetic basis of observed life history patterns. *Evolution* 36: 1236-1250.

Robertson, A. 1959. Experimental design in the evaluation of genetic parameters. *Biometrics* 15: 219-226.

Roeder, K. D. 1963. *Nerve cells and insect behavior.* Cambridge: Harvard Univ. Press.

Rose, M. R., and B. Charlesworth. 1981. Genetics of life-history in *Drosophila melanogaster.* I. Sib-analysis of adult females. *Genetics* 97: 173-186.

Schluter, D. 1984. Morphological and phylogenetic relations among the Darwin's finches. *Evolution* 38: 921-930.

Shaffer, H. B., and G. V. Lauder. 1985a. Aquatic prey capture in ambystomatid salamanders: patterns of variation in muscle activity. *J. Morph.* 183: 273-284.

Shaffer, H. B., and G. V. Lauder. 1985b. Patterns of variation in aquatic ambystomatid salamanders: kinematics of the feeding mechanism. *Evolution* 39: 83-92.

Smith, J. N. M. 1985. Determinants of lifetime reproductive success in the song sparrow. In *Reproductive success,* ed. T. Clutton-Brock. Chicago: Univ. of Chicago Press.

Stearns, S. C. 1983. The genetic basis of differences in life-history traits among six populations of mosquitofish (*Gambusia affinis*) that shared ancestors in 1905. *Evolution* 37: 618-627.

Swain, D. P., and C. C. Lindsey. 1984. Selective predation for vertebral number of young sticklebacks, *Gasterosteus aculeatus. Can. J. Fish. Aquat. Sci.* 41: 1231-1233.

Tinkle, D. W. 1967. The life and demography of the side-blotched lizard, *Uta stansburiana. Misc. Publ. Mus. Zool. Univ. Mich.* 132: 1-182.

Travis, J. 1983. Variation in development patterns of larval anurans in temporary ponds. I. Persistent variation within a *Hyla gratiosa* population. *Evolution* 37: 496-512.

Van Berkum, F. H. and J. S. Tsuji. 1985. Among-family differences in sprint speed of hatchling *Sceloporus occidentalis.* Manuscript.

Van Valen, L. 1963. Selection in natural populations: Human fingerprints. *Nature* 200: 1237-1238.

Voris, H. K. 1975. Dermal scale-vertebra relationships in sea snakes (Hydrophidae) *Copeia* 1975: 746-755.

Webb, P. W. 1984. Body form, locomotion and foraging in aquatic vertebrates. *Am. Zool.* 24: 107-120.

Wecker, S. C. 1963. The role of early experience in habitat selection by the prairie deermouse, *Peromyscus maniculatus bairdi. Ecol. Monogr.* 33: 307-325.

Weinberg, W. 1909. Uber Vererbungsgesetze beim Menschen. II. Spezieller Teil. Allgemeine Losung des Problems der Wirkung

der Panmixie bei ein facter Vermischung und alternativer
Vererbung. *Z. fur Induk. Abstamm. Vererbunstehre* 2: 276-330.

Werner, E. E., and J. F. Gilliam. 1984. The ontogenetic niche and
species interactions in size-structured populations. *Ann. Rev.
Ecol. Syst.* 15: 393-425.

Wright, S. 1921. Systems of mating. *Genetics* 6: 111-178.

11   Commentary and Conclusion

Martin E. Feder and George V. Lauder

1. The Need for Interdisciplinary Studies of Predator-Prey Relationships

Predation (in its broadest sense) affects all living things. All heterotrophs require energy for maintenance, growth, and reproduction, and all obtain this energy by degrading the chemical bonds that form the bodies of other organisms. All autotrophs, the ultimate living source of this energy, and all heterotrophs are accordingly at risk. Energy flow is a ubiquitous feature of all ecosystems. Thus, all organisms face constraints (or opportunities) that are related to predation. At the heart of most studies of predator-prey relationships lies a central question: Are there any general principles that characterize organisms' responses to these constraints and opportunities, or are the responses so diverse that no clear generalizations emerge? The clearest generalizations appear to stem from studies of ecosystems, communities, and populations. We can readily appreciate, for example, density-dependent and density-independent factors that limit both the availability of food and the opportunity to obtain it, relationships between the amount of competition and the species diversity of a community, and the general implications of the Lotka-Volterra relationship for predator and prey populations. Many of the mechanistic explanations for these generalizations remain elusive, however, and the breadth of such generalizations decays steadily with decreasing levels of organization. All organisms appear to have some response to forestall being eaten by others, but the diversity of such responses is truly astounding. All

180

heterotrophs have some mechanism by which they "perceive" their food, capture it, and process it, but natural selection and other processes have shaped these mechanisms so dramatically that most comparison is strained. This is not to say that generalizations on this level are non-existent, only that the generalization process itself becomes increasingly difficult because of the obvious diversity.

We are reminded of the parable in which several blind men cannot achieve a general picture of an elephant when each examines a different part. This is due to the diversity of the elephant's parts, but also to the limitations of each man's grasp. Perhaps touch is intrinsically a poor way to examine elephants. Likewise, our understanding of the "elephant" of predator-prey relationships is limited by both the immensity of the subject and the tools we bring to bear upon it. We have structured the symposium and this volume to concentrate upon the latter subject: tools, approaches, and methods of analysis. We began this volume by posing several questions: "What is the utility of each of a variety of approaches to understanding predator-prey relationships?" "What sorts of questions is each approach best suited to answer, and what sorts of questions is each ill-suited to answer?" "What are the major research problems that remain to be addressed from the perspective of each particular approach, and how can the different approaches best be integrated into a common general framework for studying predator-prey relationships?" We would like to end this volume with a commentary that addresses these questions in a general sense. Where do we go from here?

The universality of predation as a biological phenomenon and the diversity of its manifestations daunt most attempts at generalization. A complete understanding of predator-prey relationships is beyond the grasp of any one person. Consider, for example, the multiplicity of mechanistic elements involved in but a single predator-prey interaction: properties of the receptors with which predator and prey detect one another, the neural substrates that govern the processing of this information and initiation of appropiate responses, the number and contractile properties of the muscle fibers that potentiate movement, the metabolic processes that power movement, the effects of noxious or toxic substances that deter predators and subdue prey, the biochemical pathways that yield these substances, the biomechanical properties of structures that subdue prey or frustrate predation, the processing of ingested prey, and the repair of damage incurred in an encounter, to name just a few. Still other "evolutionary" elements are involved: the energetic and temporal cost of a predatory attempt, the risk of foraging, the behavioral optimization of foraging tactics and anti-predatory tactics in diverse environments, the functional and numerical responses to changes in prey abundance, predator effects upon prey populations and vice versa, and ultimately the evolution of optimal foraging tactics and optimal predator deterrence. Expertise in any one of these elements is potentially a life's work, and yet a grasp of all of these elements is crucial to a synthetic understanding of predator-prey relationships. Multiply this fact by the number and diversity of predator-prey relationships and the product seems overwhelming. Where do you begin? Which elements do you pursue, and which do you ignore? One obviously cannot understand all of these elements; one has to pick and choose. To a large extent, the portrait of predator-prey relationships that emerges will reflect the elements that are included in the analysis. *Our major thesis here is that the choice of elements for study has historically often been too narrow, and that greater breadth would improve our progress in understanding predator-prey relationships.*

Lack of sufficient breadth in studies of predator-prey relationships surely is understandable. One obvious barrier to greater breadth is the difficulty of keeping abreast of more than a small fraction of developments in this (or any other) large field (Bartholomew, 1982). Another, especially in the more reductionistic aspects of the field, is the time, effort, and funds that must be invested in the state-of-the-art practice of more than a single technique or approach. A third barrier is the jargon and overspecialization that retard communication among sub-disciplines. The last of these is perhaps the simplest to deal with, as we have attempted to demonstrate (although with varying success) in the course of this symposium and volume. As far as we can determine, no comparably diverse symposium on predator-prey relationships has been held before. To be sure, a diverse symposium does not guarantee any communication among workers in various sub-disciplines, especially if symposium presentations are framed as highly technical summaries of current research findings. Thus, a first (and important) step is an explicit attempt to communicate research findings to workers in allied but distinct fields, and to avoid the jargon that typically isolates sub-disciplines from one another. This volume largely stems from the contributors' willingness to make their specialty comprehensible to the interested non-specialist.

2. Interdisciplinary Impact of Natural History

Given that specialists in diverse fields can communicate, several areas emerge in which cross-fertilization would be particularly beneficial to all parties. One is "natural history." As Harry Greene emphasizes in his essay, natural history has acquired an unsavory reputation. Yet, most of the contributors to this volume and many in the audience at the symposium have called for more "good" natural history studies. Why are such studies still considered necessary?

The major insights furnished by reductionistic and experimental approaches to predator-prey relationships have often been achieved by abstracting predator-prey encounters and divorcing them from the ambiguities of the field. Predator, prey, or both are induced to perform presumably naturalistic activity in a laboratory setting, frequently with substantial amounts of instrumentation present. Is this performance normal in any sense, or has the experimenter "stacked the deck" to yield behaviors that are quantifiable but unrealistic? If the latter is the case, the results may yield important insights into the neural, mechanical, muscular, biochemical, or metabolic substrates of "behavior", but tell us little about predator-prey relationships. We must know whether the processes observed in the laboratory actually occur in the field. When Paul Webb releases a minnow in front of a voracious predator, he has staged the situation so that only a fast-start response can possibly save the minnow. In the minnow's complex environment of weeds and murky water, immobility and protective coloration may be equally effective; perhaps fast starts are irrelevant. When Gerhard Roth presents hungry salamanders with symbolic prey, he is likewise asking salamanders to behave in totally unrealistic fashion. Both cases clearly represent experimental abstractions that are absolutely necessary to render the respective research problems tractable. Yet, when it comes time to generalize from the laboratory setting to the field, will the generalizations bear scrutiny? This should not be construed as a condemnation of laboratory predator-prey studies; it is entirely possible (and, indeed, likely) that fishes and salamanders behave in the field as Webb and Roth expect them to. Of course, the opposite is possible as well. It is impossible to tell without some study of natural history.

Many laboratory and field studies of predator-prey relationships are structured about optimality arguments (e.g., Townsend and Calow, 1981). In other words, an important working assumption of most predator-prey studies is that natural selection has acted to optimize some aspect of organismal performance, be it foraging behavior per se, the design of jaws, the metabolic economy of foraging, or the protective coloration of potential prey. To be sure, predators and potential prey have evolved as if optimality were closely linked to fitness. Yet, seldom has any real penalty of non-optimality been *quantified* in a naturalistic context. One can compare a high-performance sports car and a budget sub-compact automobile in the "laboratory", conclusively demonstrate the superior design features of the former, and envision hypothetical situations in which these features may be valuable. Chances are, however, that on crowded city streets (i.e., "in the field") both kinds of vehicles will deliver you to your destination in the same amount of time. Is the same true of optimal and non-optimal behaviors, morphologies, physiologies, etc? Surprisingly few studies have considered this question, and their conclusions are mixed. On one hand, some predators clearly wander around in lackadaisical fashion as if a foraging strategy were unimportant (e.g., snake foraging as described by Harry Greene), some prey are captured merely because they happen to be in the wrong place at the wrong time (Feder, 1983), and experimental removal of supposed anti-predator adaptations may have little effect on survivorship (Silberglied et al., 1980). On the other hand, the recent popularity of field measurements of fitness and the explicit quantification of "selection gradients" (Lande and Arnold, 1983; Chapter 10) suggests that a measurable decrement in fitness may attend any deviation from a "most fit" (i.e., optimal) morphology. We have yet to resolve how often optimal morphologies or behaviors are actually necessary to avoid a decrement in fitness, and how often adequate but suboptimal traits will suffice. This information is obviously crucial to addressing the adequacy of optimality criteria, which are typically drawn from engineering, economics, or game theory rather than from biology. "Good" natural history will help yield this information.

Despite the voluminous literature of natural history, the "good" natural history addressing the issues outlined above has seldom been attempted. We specify '"good" natural history' because far too much natural history consists of raw and disorganized observations and anecdotes, which are of little value. Precise and detailed accounts are needed of behavior and environment during predator-prey encounters in the field: light levels, current speeds, temperatures, distances between predator and prey, and velocities, forces, accelerations, and metabolic rates during encounters, for example. Equally important is information on the frequency of prey encounters, predator encounters, and foraging movements. Long-term studies of numerous animals are needed to characterize individual variation. Studies structured about specific hypotheses (as opposed to unfocused recordings of all observable phenomena) would be most welcome. In his chapter, Harry Greene explores some of the reasons why few such natural history studies are undertaken. To this list we add the frequent lack of precise instrumentation that is adequate for use in the field, discomfort of field work, and the simple inaccessibility of many predator-prey interactions (e.g., a water snake pursuing a frog two meters beneath the surface of a vegetation-choked swamp). These factors make "good" natural history difficult but not impossible, as several exemplary studies demonstrate: Harry Greene's week-long observations of individual snakes in the Costa Rican rain forest (Chapter 7), Paul Sherman's (1977, 1981) summer-long observations of individual animals in a ground squirrel population in the mountains of California, and Timothy Clutton-Brock's (e.g., Clutton-Brock et al., 1982) characterization of "lifetime fitness" in red deer. Even these extraordinarily labor-intensive studies have not provided all of the information needed to address the questions raised above. Although more "good" natural history information would clearly be valuable in several respects, we are not optimistic about its imminent appearance.

3. Guidelines for Interdisciplinary Studies

Developments in "functional" and "evolutionary" aspects of predator-prey relationships have largely proceeded along independent lines. There is a large, robust, clearly articulated "evolutionary" avenue of predator-prey studies within the larger disciplines of ethology, ecology, and evolutionary biology. This avenue has produced important and readily identifiable theoretical contributions concerning interspecific competition, optimal foraging and other optimality considerations, population regulation, regulation of species diversity, quantitative genetics, natural selection, and coevolution (Futuyma and Slatkin, 1983; Taylor, 1984). Observation is an important research tool; experimentation, when used, is typically non-invasive. Predator-prey studies are valued for their own sake.

"Functional" studies are very different in each of these respects. Most "functional" biologists who study predator-prey relationships (e.g., Gans, Webb, Roth, Bennett) consider themselves as morphologists and physiologists first, and use predator-prey interactions as a convenient system with which to study morphology and physiology. Their approach is typically experimental and quite often invasive.

It is not surprising that the day-to-day research of functional and evolutionary biologists is largely separate, but the lack of true interaction among these two modes of inquiry is nonetheless disconcerting. Interaction would clearly benefit each approach in several identifiable ways. Those using the "evolutionary approach" would benefit from some quantification of those abstract elements that populate their models and hypotheses: cost of prey handling, predator or prey detection thresholds, cost of foraging, etc. Many of their optimality arguments implicitly or explicitly assume that the phenotype is subject only to the constraints built into their models. In many cases the "functional biologists" either have already quantified or can quantify these variables and constraints. In particular, "functional biologists" can provide important insights into the physiological, morphological, mechanical, and biochemical constraints upon variation in the phenotype. The "functional biologists", on the other hand, typically view the world as an ordered system in which morphologies and physiologies either work or fail on account of readily identifiable factors. This view often has little appreciation for variability, either among individuals or among microhabitats, and little sense for the probabilistic nature of predator-prey encounters in the field. "Evolutionary" students of predator-prey relationships could do a valuable service for their "functional" colleagues by sharing their insights into variability and probability. We offer as evidence for this point the results of two research programs that integrate "functional" and "evolutionary" approaches to predator-prey relationships, one of which involves broadly based collaboration.

Bernd Heinrich's examination of foraging in insects (principally bumblebees) has ranged from biochemical specializations in individual cells and tissues to foraging patterns of individual insects to the operation of the hive as a social entity to the implications of foraging for community structure. Each of these levels of approach has contributed substantially to Heinrich's understanding of other levels, an understanding that would have been lost had Heinrich's research not been interdisciplinary in nature. As described by Heinrich (1979), the physiological and morphological substrates of long-distance foraging (i.e., flight) are such that muscle temperature must remain within narrow limits for effective flight. This insight stems from "functional" studies of cells, tissues, and whole organisms in a laboratory setting. Because of this constraint, bees must either confine their foraging to moderate temperatures or use costly thermoregulatory mechanisms to maintain the correct temperature for flight. The enormous energetic cost of flight is a consequence of the physiological mechanisms used to sustain it and has important implications for the "natural economy" of bees. On warm days, bees can reap an energetic profit by visiting flowers containing a relatively small amount of food; on cool days, thermoregulation increases the cost of foraging and hence constrains bees either to visit plant species whose flowers contain much more food or minimize flight during foraging. Hence, a plant species that "forages" for insect pollinators must balance its energetic inducement for insect visits against seasonal variation in the cost of thermoregulation. This insight stems ultimately from the laboratory studies. Indeed, one can relate the sequence of flowering of species in a plant community, and the morphology of each species' flowers, and some aspects of plant community structure to this interplay among the energetic costs and climatic restrictions upon the foraging of the pollinators. In turn, recognition of the climatic and plant-imposed constraints upon foraging lead back to the laboratory, where biochemical, physiological, and morphological specializations that increase the effectiveness of thermoregulation have been discovered.

Such breadth is often difficult to achieve in a single laboratory, but equally rewarding results can emerge from a collaborative study involving several investigators. A good example is an ongoing study of lizards in the Kalahari desert, described by Raymond Huey and Albert Bennett in this volume. The original research question posed by Eric Pianka concerned whether desert lizard communities in diverse geographic localities had similar structures. Pianka and Huey began to answer this question through a field study structured about competition theory and community ecology, in which resource utilization (i.e., predator-prey relationships) was a conspicuous component. One finding that emerged from behavioral and natural history observations was of species pairs that differed in foraging mode: sit-and-wait vs. widely-foraging. This observation then raised the question of whether the behavioral differences were purely voluntary or reflected physiological constraints on movement. Addition of two comparative physiologists (Albert Bennett and Henry John-Alder) to the program answered this question: some (but not all) metabolic variables and laboratory measures of performance clearly indicate a constraint. Moreover, the combination of field metabolic measurements (with the aid of Kenneth Nagy) and observations of foraging in free-ranging lizards was able to establish the greater profitability of wide foraging with respect to sit-and-wait predation. Thus, a question in theoretical ecology led to a natural history study that documented significant behavioral differences, which became the subject of a physiological observation. These physiological studies will in turn generate characters to be used in a future examination of the evolution of foraging modes in lizards, which will also incorporate behavioral and demographic data. Here again we see how progress in one sub-discipline may foster progress in another, sometimes in unforseen ways.

A colleague of ours has defined the interactive distance of biologists as "the farthest one can carry a cup of coffee without it getting cold." In a very real sense, physical proximity is a necessary condition for ongoing collaboration among biologists in diverse specializations. The examples cited above notwithstanding, many factors in the academic environment seem to work against cross-fertilization. Faculty are increasingly being segregated into academic departments of evolutionary biology, organismal biology, cell biology, and molecular biology instead of being combined into departments that span diverse levels of biological organization. The structure of research funding makes it most practical to develop a research program that does one thing very well; to try to do several things less well invites early retirement. What, then, can be done to promote collaborative research into predator-prey relationships? We realistically see little that can be done about the increasing specialization of academic departments. Any non-specific exhortation to increased collaboration with no additional incentive is likely to be ineffective. Perhaps a careful consideration of appropriate incentives by funding agencies (e.g., the National Science Foundation) would yield the greatest progress in this area. For example, reservation of a small but well-publicized number of grants for support of truly interdisciplinary studies of predator-prey relationships might well spur increased breadth in research programs.

4. Conclusion

As stated in the opening sentence of Chapter 1, this is not a typical symposium volume. We have attempted to stress conceptual and practical issues in the *analysis* of predator-prey relationships, as opposed to the data themselves. Each of the chapters discusses questions that will be foci of research in the coming years. We hope that the speculation, general discussion, integration among chapters, and the relatively small quantity of data presented will allow those interested in predator-prey relationships an entree into areas other than their own specialty. Above all, we hope that this volume will stimulate researchers to think about connections among their own area and other components of predator-prey relationships, and will help foster the kind of interdisciplinary interactions that are so essential to the development of a general perspective on predators and their prey.

Acknowledgments

We thank the two anonymous reviewers of this chapter (and, indeed, the entire volume) for their helpful comments. Preparation of the manuscript was supported by NSF Grant BSR83-20671.

References

Bartholomew, G. A. 1982. Scientific innovation and creativity: a zoologist's point of view. *Am. Zool.* 22: 227-235.

Clutton-Brock, T. H., F. E. Guinness, and S. D. Albon. 1982. *Red deer: the behavior and ecology of two sexes.* Chicago: Univ. of Chicago Press.

Feder, M. E. 1983. The relation of air breathing and locomotion to predation on tadpoles, *Rana berlandieri*, by turtles. *Physiol. Zool.* 56: 522-531.

Futuyma, D. J., and M. Slatkin. 1983. *Coevolution.* Sunderland, Mass.: Sinauer.

Heinrich, B. 1979. *Bumblebee economics.* Cambridge: Harvard Univ. Press.

Lande, R., and S. J. Arnold. 1983. The measurement of selection on correlated characters. *Evolution* 37: 1210-1226.

Sherman, P. W. 1977. Nepotism and the evolution of alarm calls. *Science* 197: 1246-1253.

Sherman, P. W. 1981. Reproductive competition and infanticide in Belding's ground squirrels and other animals. In *Natural selection and social behavior: recent research and new theory,* ed. R. D. Alexander and D. W. Tinkle, pp. 311-331. New York: Chiron Press.

Silberglied, R. E., A. Aiello, and D. M. Windsor. 1980. Disruptive coloration in butterflies: lack of support in *Anartia fatima*. *Science* 209: 617-619.

Taylor, R. J. 1984. *Predation.* New York: Chapman and Hall.

Townsend, C. R., and P. Calow. 1981. *Physiological ecology: an evolutionary approach to resource use.* Sunderland, Mass.: Sinauer.

Index

Abiotic factors, in ecological
    systems, 109
Absorbance curve, 124
Acceleration, 32
    capability, 25
    capacity for, 31
    centripetal, 36
    maximal, 27, 28
Active flight, 140
Activity
    intensity of, 70
    level, 25
    timing, 112
Adaptation, 9, 10, 22, 102, 110, 157
Aeoliscus, 113
Aerobic metabolism, 36, 70, 76
Aerobic scope, 70, 89
Aggregations, 115
Ambloplites rupestris, 31
Ambystoma, 112
Amphibians, 42-65
Anaerobic capacity, 71
Anaerobic metabolism, 36, 70, 71,
    74, 75, 76
Analysis of variance, 163, 165
Anatomy, 100
Anolis, 113
Anti-predator defenses, 109-132

Aposematic coloration, 114, 131
Apostatic selection, 112
Approach, stage of predation, 110
Autecology, 99, 101
Autotrophs, 180
Avoidance learning, 114
Avoidance behavior, 137, 141

Background matching, 126
Batesian mimicry, 114
Behavior, 8, 26, 34, 35, 38, 90, 158
    escape, 126
    neural mechanisms of, 44
    of predators, 82
    of prey fishes, 135-152
Behavioral energetics, 73, 77
Behavioral responses, constraints
    on, 69
Benthic habitat, 123
Biological organization, 84
Biological role, 103, 157
    also see Role
Biomechanics, 10
    research, 36
Biomechanical studies, 39
Biotope, 7
Bitis caudalis, 92-93
Bitis peringueyi, 103

Bolitoglossini, 65
Bothrops asper, 102
Breeding designs, 168
Brightness, 124
Bufo, 52, 53, 58, 64
Bufo bufo, 46
Bufo typhonius, 116
Bushmaster, 103, 104

C-start, 28
    acceleration, 30
Carbon dioxide production, 72
Catfish, 116
Cause and effect, 126
Ceratophrys, 116
Chamaeleo, 113
Chemosensory senses, 129
Chlamydosaurus, 114
Chordates, 74
Chromatophores, contractile, 130
Cicadas, 74
Cichlasoma nigrofasciatum, 144
Cichlasoma rostratum, 144
Cichlid fishes, 144
Cinefluoroscopy, 12, 14
Cinematography, 12, 14
Cnemidophorus tigris, 102
Coachwhip, 102
Coevolution, 185
Color, 119
    brightness of, 124, 125
    discrimination of, 130
    conspicuous, 114
    diversity, 123
    pattern, 115, 122, 127, 128
    cryptic pattern of, 119
    spot, 126
Command neurons, 53
Comparative
    approach, 21
    biology, 9
    method, 82
    studies, 8, 37
Competition
    presumed, 152
    theory, 150

Computer simulations, 89
Confusion, by sensory effects, 113
Constraints, 129
    in models, 186
Consumption, 117
    stage of predation, 110
Correlated characters, 160
Cost, 70, 73
    of activity, 77
    of searching and
      handling, 103, 104
Cost-benefit
    calculations, 35, 136, 142
    relations, 7
    solutions, 11
Costs and benefits, of behaviors,
    141
Crepuscular habitat, 123
Crocodiles, 76
Crotalus, 114
Crotalus cerastes, 102
Crotalus tigris, 102
Crustaceans, 74
Crypsis, 112-113, 115, 127, 128,
    131
    proper, 113
Ctenosaura, 116
Ctenosaura similis, 161
Culaea inconstans, 140
Cyclops, 37
Cyrtocara eucinostomus, 137

Dascyllus aruanus, 141
Davis, D. Dwight, 22
Defense mechanisms, multiple,
    127
Dendrobatidae, 114
Density-dependent predation, 112
Depth perception, 47, 59
Deserts, 22
Desmognathus, 114
Detection, stage of predation, 110
Developmental constraints, 129
Diademichthys, 114
Differential survival, 118
Disruptive coloration, 127

Diurnal eye, 145
Diving, 76
Doradid catfish, 116

Ecological studies, 42
Ecology, of predation, 127
        of predators, 82
Efficiency, 7
Elapidae, 114
Electric senses, 129
Electromyography, 12, 13
Eleotrids, 144
Eleutherodactylus, 112, 113
Elgaria, 114
Energetics, 1
Energy, 6, 7, 35, 69, 180
        budget, 70
        expenditure, 69
        flux, 25
        utilization, 72
Ensatina eschscholtzi xanthoptica,
        114
Ensemble coding, 44
Escape activity, 140
Esox, 30, 31, 33, 34
Esox lucius, 30, 141
Esox masquinongy, 30
Ethological
        observations, 135
        studies, 42
Ethology, descriptive, 99, 101
Eucrypsis, 119, 130
Evolution, direction of, 91
Evolutionary morphology, 100
Exaptation, 110, 112
Exhaustion, 70
Experiments, natural vs.
        manipulative, 85
Eyes, 145

Fatigue, 76
Feature detector, concept of, 44,
        45, 53
Feeding behavior, 43, 46
        of amphibians, 44, 65
        sensorimotor system
        guiding, 45

Feeding ecology, 43
Fer de lance, 102
Field conditions, 71
Fins, folding, 115, 130
Fitness, 136, 152, 158, 172
        field measurement of, 169
        in the field, 170
Fitness maximization, 136
        theory, 144
Flight, 22
Foraging, 71
        ecology, 100
        models, 90
        strategy, 183
        theory, 103
        in insects, 186
Foraging mode, 85, 91, 187
        coexistence of, 93
        ecological correlates of,
        86, 87
        effects on diet of, 85
        effects on predation of, 92
        energetics of, 92
        evolution of, 91
        in Kalahari lizards, 83-84
        mechanistic basis of, 88
        origin vs. maintenance of,
        92
        physiological correlates
        of, 89
        sensory differences in, 86
Force plates, 12
Frogs, 74, 113
Functional
        anatomy, 126
        biologists, 186
Functional morphology, 2, 4, 6-23
        of locomotion, 26
        value of, 36
Functional morphologists, 136
Functional response, 3

Game theory, 34, 35
Games, differential, 25
Gape-limited predators, 116
Gas exchange, 71
Gasterosteidae, 140

Gasterosteus, 116
Generalist, 7
Genetic covariance, between
    performance and fitness,
    172
Gobiomorus, 145
Gobiomorus dormitor, 144
Grunts, 148

Habitat, 123
    switch, 139
Haemulidae, 146
Hearing, 129
Heart
    mass, 161
    rate, 72
Heinrich, Bernd, 186
Heritability, 8, 165, 166, 167
    estimates of, 166
    standard error of, 168
Heterotrophs, 180, 181
Home range, 151
Horned adder, 92
Horned vipers, 116
Humbug damselfish, 141
Hummingbirds, 104
Hydromantes, 52, 53, 58, 64, 65
Hydromantes italicus, 46
Hyla geographica, 131

Identification, stage of predation,
    110
Immobility, 140
Inequalities, 26
Inheritance, 159, 160, 164-166, 171
    of behavioral traits, 164
    of morphological traits,
    164
    of physiological traits, 164
Insects, 186
Intraspecific communication, 127
Iteroparous spawners, 144

Kalahari desert, 83
Katydids, 74
Kin selection, 112, 116, 131

Laboratory conditions, 71
Lachesis muta, 103
Lactate, 74, 75
Largemouth bass, 137
Lateral line senses, 129
Leaf fish, 113
Lekking ground, 137
Lepomis macrochirus, 137
Lethality, as a defense, 116-117
Life tables, 144
Lions, 25
Lizardfish, resin-coated, 152
Lizards, 113, 114
    varanid, 104
Locomotion, 2, 24-39, 73
    on a treadmill, 73
    performance during,
    161
Locomotor
    biomechanics, 30, 34
    capacity, 8
Longnosed snake, 102
Looming image, 33
Loricariid catfishes, 137
Lotka-Volterra relationship, 180

Masquerade, 113, 130
Masticophis flagellum, 102
Maternal effects, 166
Mauthner cells, 28
Mertensian mimicry, 117
Mesopic conditions, 145
Metabolic
    cost, telemetering, 72
    rate, 12, 72
    systems, 69
Metabolism, 36
    components of, 89
    organismal, 4
Micropterus dolomieui, 31
Micropterus salmoides, 137
Midwater habitat, 123
Migration, 146, 148
Mimicry, 114, 116
Models, of predators, 152
Molluscs, 74

Moloch, 113, 116
Moloch horridus, 103
Monocirrhus, 113
Morphology, 90, 158, 159
Moths, 74
Movie film, 25, 34, 35
Mullerian mimicry, 114, 117
Multidisciplinary approaches,
    importance of, 90
Multiple predators, effects of, 128
Muscle, 12, 13, 16, 17, 18
    fibers, 181
    fiber architecture, 12
    histochemistry, 12
    length-tension relation, 16
    mass of leg, 161
    packing, 16, 20
    pinnation of, 20
    sarcomeres, 16, 17
Muskellunge, 30
Myotome, 28

Natural experiment, 85
Natural history, 2, 99-105, 150,
    182, 183, 187
Natural selection, 109, 110, 118,
    136, 158, 159, 181, 185
Neetroplus nematopus, 144
Neural networks, 43
Neurobiological studies, 42
Neurobiologists, 136
Neuroethology, 43
Nocturnal eye, 145
Non-sustainable behaviors, 70
Northern pike, 30, 141
Notophthalmus, 114
Noturus, 116

Object recognition, neural
    mechanisms of, 44
Ontogeny, 160, 163-164, 171
Optic tectum, 49
Optimal foraging, 182, 185
Optimality
    arguments, 183
    criteria, 184

Optimization, 136
Optimum
    behavior, 24, 26, 28, 30,
    34, 39
    form, 38
    strike tactics, 33
Oxygen consumption, 70, 72, 73,
    76
    maximal, 70, 161
    rate of, 71
Oxygen transport, 89

Palm tree, 103
Parasites, 131
Parental care, 142, 144
Parrotfishes, 139
Parsimony, 91
Patch size, of spots, 126
Pattern recognition, by predators,
    130
Performance, 37, 158, 163, 164,
    167
    during ontogeny, 171
    in the field, 170
Phenotype, 8, 9, 10, 11, 21, 22
    mother's, 166
Phosphate stores, high energy, 75
Phosphocreatine, 75
Photopic conditions, 145
Phrynosoma, 113, 116
Phylogenetic
    accidents, 117
    analysis, 91
    constraints, 129, 130
    control, importance of,
    85-86
    correlates, of spawning
    stupor, 143
    perspective, 9
Phylogeny, 83
Physiology, 90, 158
Pine martins, 104
Piscivores, 136
Plasticity, in prey response, 152
Plethodon, 114
Plethodon jordani, 114

Plethodontini, 65
Pleurodeles, 116
Poecilia reticulata, 112, 122
Poisonous species, 117
Polycentrus, 113
Polygenic traits, 160, 165
Polymorphic prey, 128
Population biology, 2, 101
Population ecologists, 136
Preadaptations, 110
Predation
       cycle, 26
       event, components of, 110
       stages of, 109
Predator avoidance, 136
Pretectum, 56
Prey
       camouflaged, 25
       detectors, 45
       distribution, 137
       handling, 20
Prey recognition, neural
       mechanisms of, 43, 44
       neurons involved in, 60,
       62, 64
Private channels, 130, 131
Propulsive segment, 28
Protoadaptations, 110
Pseudotriton, 114

Quantitative genetics, 2, 165, 185
       theory of, 160

Radiotelemetry, 103
Rainbow trout, 31
Rana, 52, 53, 64, 65
Rana esculenta, 46
Rana pipiens, 56, 59
Rana temporaria, 46
Rarity, 112
       apparent, 112
Reaction forces, 28
Recognition module, 64, 65,
       concept of, 56-57
Reflectance, 119, 122

Refuge seeking, 140
Repeatability, 164
Reproduction, 142
Reproductive behavior, 136
Reproductive success, lifetime, 169
Response threshold, 33
Retina, 47
       ganglion cells, 47, 49
Rhinocheilus lecontei, 102
Ribs, 129
Rivulus hartii, 115
Rock bass, 31
Role, 8, 9, 11, 21
Running speed, 70

S-start, 30
       acceleration during an, 31
Sacramento perch, 25
Salamanders, 74, 76, 112
       plethodontid, 42
Salamandra, 52, 53, 64, 116
Salamandra salamandra, 46
Salamandridae, 114
Salmo gairdneri, 31
Sanddiver lizardfish, 146, 148
Scaling, 10
Scaridae, 139
Schooling, 139, 146
Schools, of fish, 115, 152, 153
Scorpaenid fishes, 116
Scotopic conditions, 145
Sea turtles, 72
Selection, 136, 160, 170
       coefficient, 163
       correlational, 161-162
       directional, 161-162
       measuring, 160
       sexual, 127, 130
       stabilizing, 157, 161-162,
       173
       surface, 161
Semelparous spawners, 144
Sense organs, adaptive evolution
       of, 43
Sensorimotor integration, 60

Sensory
  ability, 123
  biology, of predation, 127
  mode, 120, 129
  physiology, 2, 4, 8
Sidewinder, 102
Signal to noise ratio, 112
Sit-and-wait
  predators, 32, 84, 85
  lizards, 86
Skin, of reptiles, 102-103
Smallmouth bass, 31
Snakes, 76
Spawning
  aggregations, 142
  stupor, 142-144
Specialist, 7
Spectral sensitivity curve, 123
Spiders, 74
Spines, as a defense, 116, 140
Startle behavior, 37, 115
Startle response, 28, 36
Sticklebacks, 140
Strain gauge recordings, 12
Strength, as a defense, 116
Strike, 27, 28, 30, 32, 34, 35, 47
  angle, 30
  paths, 37
  success, 33
  trajectory, 29
Structure-function association, 14
Subjugation, stage of predation, 110
Sunfish, 137, 139
Surface habitat, 123
Survivorship curves, 144
Sustainable behaviors, energetics of, 70
Syngnathus, 113
Synodus intermedius, 146
Systematic biologists, 105
Systematic biology, 101

Tadpoles, 75, 131
Tail loss, 128
Target, 29

Taricha, 114
Taricha torosa, 114
Telemetry, 72
Termites, 87
Thamnophis, 126
Thecadactylus, 113
Thermoregulation, 127
Tiger musky, 30
Toughness, as a defense, 116
Toxins, 116
Transplant experiments, 152
Treadmill, 73, 89
Tree frogs, 112
Trophic level, 6
Turtles, 76
Twilight
  activity, 136
  changeover, 145
Tylotriton, 116

Unken display, 115

Variation, 160
  among individuals, 3, 86, 186
  among litters, 166
  and selection, 169
  genetic, 118
  geographic, 128, 129
  in brightness, 123
  in defense mechanisms, 118
Vertebrae, 167
Video recordings, 34, 35
Video tape, 25
Vipers, tropical, 102
Vision, 33, 112, 129
Visual
  acuity, 126, 130
  crypsis, 122
  signals, 112, 114
  system, amphibian, 47
Vulnerability, in young, 153

Warning coloration, 131
Water, doubly-labelled, 72

Welfia georgii, 103
Whiptail lizard, western, 102
Wide-searching species, 32

Widely-foraging
    predators, 84, 85
    lizards, 86
Worm detector, 53